LAND ROVER DEFENDER, 90 AND 110 RANGE

30 Years of the Coil-Sprung 4 x 4 Models

OTHER TITLES IN THE CROWOOD AUTOCLASSICS SERIES

AC COBRA Brian Laban

ALFA ROMEO 916 GTV AND SPIDER Robert Foskett

ALFA ROMEO COUPÉS 1954–1989 Graham Robson

ALFA ROMEO SPIDER John Tipler

ASTON MARTIN DB4, DB5 & DB6 Jonathan Wood

ASTON MARTIN DB7 Andrew Noakes

ASTON MARTIN V8 William Presland

AUDI QUATTRO Laurence Meredith

AUSTIN HEALEY Graham Robson

BMW 3 SERIES James Taylor

BMW 5 SERIES James Taylor

CITROËN DS SERIES John Pressnell

FORD CAPRI Graham Robson

FORD ESCORT RS Graham Robson

JAGUAR E-TYPE Jonathan Wood

JAGUAR XJ-S Graham Robson

JAGUAR XK8 Graham Robson

JENSEN INTERCEPTOR John Tipler

JOWETT JAVELIN AND JUPITER Geoff McAuley & Edmund Nankivell

LAMBORGHINI COUNTACH Peter Dron

LANCIA INTEGRALE Peter Collins

LANCIA SPORTING COUPÉS Brian Long

LOTUS & CATERHAM SEVEN John Tipler

LOTUS ELISE John Tipler

LOTUS ESPRIT Jeremy Walton

MGA David G. Styles

MGB Brian Laban

MGF AND TF David Knowles

MG T-SERIES Graham Robson

MASERATI ROAD CARS John Price-Williams

MERCEDES-BENZ CARS OF THE 1990S James Taylor

MERCEDES-BENZ SALOONS OF THE 1960s AND 70s Laurence Meredith

MERCEDES SL SERIES Andrew Noakes

MORGAN THREE-WHEELER Peter Miller

MORGAN 4-4 Michael Palmer

ROVER P5 & P5B James Taylor

SAAB 99 & 900 Lance Cole

SUBARU IMPREZA WRX AND WRX STI James Taylor

SUNBEAM ALPINE AND TIGER Graham Robson

TRIUMPH SPITFIRE & GT6 James Taylor

TRIUMPH TR7 David Knowles

VOLKSWAGEN TRANSPORTER Laurence Meredith

VOLKSWAGEN GOLF GTI James Richardson

VOLVO P1800 David G. Styles

LAND ROVER DEFENDER, 90 AND 110 RANGE

30 Years of the Coil-Sprung 4 x 4 Models

James Taylor

THE CROWOOD PRESS

First published in 2013 by
The Crowood Press Ltd
Ramsbury, Marlborough
Wiltshire SN8 2HR

www.crowood.com

British Library Cataloguing-in-Publication Data
A catalogue record for this book is available from the British Library.

ISBN 978 1 84797 453 2

Typeset by Shane O'Dwyer, Swindon, Wiltshire

Printed and bound in India by Replika Press Pvt Ltd

CONTENTS

ACKNOWLEDGEMENTS

Anybody who thinks that all of the coil-sprung Land Rovers are pretty much of a muchness will be surprised by the story contained within this book. When I came to pull it all together, it surprised me, too. In thirty years, the coil-sprung models have embraced an astonishing number of variants – and that is without considering the specialist conversions, which are varied enough to deserve a book of their own one day.

The essence of this book has come from my own collection of material, filed away over a period of more than thirty years in the belief that it would eventually amount to something. However, dozens, possibly hundreds, of people have added little snippets of information – and sometimes very large ones – to make the story hang together. A great deal of it has come in through my work for various Land Rover magazines: *Land Rover Owner* from 1989 to 1999; *Land Rover Monthly* from 1999 to 2000; *Land Rover Enthusiast* (as Editor and, latterly, co-owner) from 2000 to 2010; *Land Rover Monthly* again from 2010 to 2012; and, most recently, *Land Rover Owner* again. I'm grateful to all those who sent material for my long-running *Roverphile* column that has found its way into this book, and also to my colleagues on those magazines who have added their own information to the sum of my knowledge.

I have also drawn on the memories of very many Land Rover engineers and other employees, who have patiently put up with my questions or dug out ancient photographs to illustrate a point. So special thanks here must go to George Adams, Colin Andrews, Bill Baker, Tom Barton, John Bilton, Keith Bird, Mike Broadhead, Roger Conway, Larry Court, Mike Gould, Ian Jones, Geof Miller, Bill Morris, Tony Poole, Dave Saddington, Bob Seager and Les Wilkins. Equal thanks go to a smaller number of people who have asked for their names not to appear in print – and of course to anybody I should have acknowledged but have omitted from this list.

I have been very fortunate to have extensive access to the legendary Dunsfold Collection of Land Rovers, established by Brian Bashall and now run by his son Philip. Jan Valentino and Richard Bacchus in the BMIHT archives at Gaydon have been enormously helpful, and I am particularly pleased to say a public thank you to my friend Roger Crathorne, who has helped out in all kinds of ways on this project.

JAMES TAYLOR
Oxfordshire, June 2012

CHAPTER ONE

LIFE AND TIMES OF THE COIL-SPRUNG LAND ROVERS

By 1983, when the One Ten was introduced as the first coil-sprung Land Rover, the Land Rover marque had existed for thirty-five years. In all that time, its suspension had resolutely depended on leaf springs and the crucial novelty of the new model was that it had all-round coil springs. That was far from its only new feature of course, but it represented a defining moment in the history of the Land Rover.

During those thirty-five years, the Land Rover had made a place for itself in the motoring world. It had started out as a light commercial vehicle and despite the introduction of personnel-carrying station-wagon variants, it had remained essentially a light commercial. No matter that enthusiasts were wont to buy time-expired examples and turn them into competitive vehicles for off-road motor sport. No matter that Land Rovers were the first choice of pioneers and explorers from all round the world. In an August 1982 feature on buying second-hand examples, the magazine *What Car?* summed the position up quite simply: 'The Land Rover,' it said, 'is first and foremost a workhorse.'

Thirty years later, the Land Rover, renamed a Land Rover Defender in 1990 to give it a clearer identity within an expanding product range, is still primarily a workhorse. But that definition has become considerably more blurred as the people-carrying Station Wagon models have taken an increasing share of the business. A Land Rover Defender is now also a vehicle that symbolizes adventure and personal freedom. It can even be chic and it most definitely reflects the owner's view of his or her own image. Whereas the mainstream motoring press only rarely mentioned Land Rover developments in the early 1980s, by 2010 they were hot news, worthy of as many column inches as developments on everyday saloon cars.

These changes in public perception had their own effect on the development of the Land Rover; a manufacturer, after all, has to deliver products that people actually want to buy. Yet public perception was far from the only influence behind the changes to the Land Rover after 1983.

During the 1980s, four-wheel-drive models suddenly became fashionable as family transport, resulting in more and more manufacturers introducing products that threatened Land Rover sales. So Land Rover developments were driven by the need to keep ahead of the game, or, at the very least, up with it. Then there were changing regulations in the myriad world markets where the vehicles were sold – changes that affected such things as lighting, exhaust emissions and safety. Development was also affected by the backlash against four-wheel drives in general, engendered by those with environmental concerns who believed that the heavy fuel consumption typical of Land Rovers and their ilk was contrary to the best interests of the planet.

Yet some things did not change in the period of the coil-sprung models. Military demand worldwide for Land Rovers as light front-line vehicles barely fluctuated, despite the increasing popularity of purpose-built multirole machines like the Humvee. Police, ambulance and fire services continued to buy Land Rovers for special-duty tasks. Explorers, adventurers, conservationists and environmentalists all continued to buy Land Rovers to take them to places that would otherwise be inaccessible. Off-road driving enthusiasts still bought older models to turn into off-road competition or adventure machines – and often spent very large sums of money converting and equipping brand-new examples as well.

Ironically, as the new One Ten was announced in 1983, so one of its key overseas markets had started to collapse. Sub-Saharan Africa had always been a big market for Land Rovers, in particular because large fleets were bought by Government agencies and these fleets were renewed on a regular basis. Changes in UK Government subsidies at the start of the

1980s severely undermined this market, with Japanese competitors being quick to move in.

Land Rover responded in two main ways. First, it switched its focus from developing markets like these to the developed markets of the West – in particular continental Europe and North America. Then, to keep the Land Rover marque alive in Africa in the hope that better times would come, it established a number of reconditioning plants where older Land Rovers could be rebuilt to as-new condition for less than the cost of a new Japanese vehicle. Gradually, sales returned to sensible levels, although they never did regain the volumes seen in the 1960s and 1970s.

These were the company's successes in the face of adversity, but Land Rover did not always get things right. A pig-headed refusal to accept that larger and more powerful engines, most notably a diesel, were an absolute necessity in some markets allowed competitors to make major inroads into long-standing Land Rover markets in the early 1980s. Without what amounted to unilateral action by some of the marque's overseas branches, the situation would have been worse.

There is also no doubt that development of the workhorse models was neglected to an extent in the 1990s. The new focus on developed markets certainly diverted resources from developing a new Defender. The models that Land Rover developed instead – the Discovery, the Freelander and the Range Rover Sport – undoubtedly did the marque a power of good, but the constant deferment of a new model to some indeterminate point in the future meant that the task of developing one became ever more difficult. When it became clear, in 2011, that work really had begun on a Defender replacement, it also became apparent that the new model would have to be very different from the old in very many ways. The world, quite simply, had moved on.

Changes of Ownership

Yet the coil-sprung Land Rovers did provide a form of reassuring continuity through all the changes of ownership that their manufacturer went through at the end of the twentieth century and the start of the twenty-first. Owners came and owners went, but the Land Rover, a Defender by then, remained a constant.

When the One Ten was announced in 1983, Land Rovers were made by Land Rover Ltd, which was a standalone operating division of the much-reviled British Leyland (BL). That change to its status had occurred as recently as 1978, when Land Rover had been established as a separate company. Before then, it had sat uncomfortably alongside Rover, Triumph and Jaguar cars in the specialist cars division of the company, because before being absorbed into BL at the end of 1966, Land Rovers had been built by the old Rover Company.

However, it was Government money that had made the One Tens possible. Their design and development was funded by a £200 million grant provided as Stage 2 of an investment scheme agreed in 1978 – and as a result these models were initially known within Land Rover as Stage 2 types.

Arguably, this was good use of the British taxpayer's money. However, the burden could not be borne forever and in 1986 the Government made moves to sell British Leyland. At that stage, the most likely buyers for Land Rover were General Motors and Ford, but public protests made the sale impossible. A management buyout attempt in 1987 was rejected; negotiations for British Aerospace to take over the whole of the Rover Group were announced in March 1988 and the deal was ratified by shareholders on 11 August. Only a nominal purchase price was paid (and this was the subject of later condemnation by European bodies), as the deal was as much a political move as anything else. The Rover cars and Land Rover sides of the company had been swiftly integrated so that no buyer could easily close down the loss-making car business and focus on the profit-making Land Rovers. That safeguarded jobs at Rover – which was important for the Government.

Nevertheless, British Aerospace ownership had been planned as a five-year caretaker arrangement; thus by late 1993 the company was looking for somebody to take over the burden of the Rover Group. Although there was a long-standing relationship between the Rover Cars side of the business and Honda in Japan, that was not exploited. Instead, British Aerospace was receptive to an approach from the German BMW company, whose interest was primarily in Land Rover. The introduction of the Discovery in 1989 had persuaded BMW to sit up and take notice, then in 1991 the German company had agreed to supply diesel engines for the forthcoming new Range Rover, due in 1994. Things became more serious towards the end of 1993 and in January 1994 the deal was done. The whole of the Rover Group became a BMW subsidiary.

BMW entered its period of ownership with enthusiasm. Over the next few years, it invested heavily in Land Rover, although most of that investment bypassed the workhorse models. Unfortunately, part of the deal with BAe had been that the German maker should also take on Rover Cars, the car-making business which was the mortal remains of all the marques once united under BL. This division made big losses, with things coming to a head in 2000 when the BMW Board refused to continue supporting it. Rover Cars and Land Rover were sold separately, Land Rover going to Ford in the summer of that year.

Ford, too, invested heavily in Land Rover, although once again the major investment bypassed the Defender range. An important change during this period was that Land Rover was amalgamated with car-maker Jaguar, also then under Ford ownership, to create a new company called Jaguar Land Rover (JLR). Then Ford's core business in the USA ran into trouble and the company chose to slim down by selling off some of its peripheral businesses. Jaguar Land Rover was among them, and was bought in 2008 by the giant Indian Tata Group. Ironically, at the time Land Rover was celebrating its sixtieth anniversary since the very first production models in 1948.

The Men in Charge: Corporate Management

Individuals at the top of the Land Rover corporate organization had an important influence on the company's products and well-being over the years. When the One Ten was brought to market in 1983, Land Rover's Managing Director was Mike Hodgkinson, who had held that position since the company's foundation in 1978. But his successor had already been appointed and from later in 1983 it was Tony Gilroy at the helm. Gilroy was notoriously forthright, but laid the foundations for Land Rover's resurgence during the 1980s, in particular overseeing the reduction in 'satellite' manufacturing plants, which had been foreseen under Hodgkinson, and the reorientation towards developed markets.

Gilroy was still at the helm of Land Rover when the Rover Cars and Land Rover businesses were reorganized as the Rover Group under Sir Graham Day. In 1987, he led a management buyout attempt, but that was rejected. When Day appointed George Simpson as Group Managing Director with the job of integrating the cars and Land Rover sides of the company into a single operation, Gilroy left. A new streamlined management structure was announced in January 1989 and, under the new regime, Land Rover's most senior figure was its Commercial Director, Chris Woodwark.

The merged Rover Cars and Land Rover operations remained so when BMW took over, but the German company put in its own top management. From early 1994, Land Rover was directly managed from Germany as part of the Rover Group, initially by Wolfgang Reitzle, then by Walter Hasselkuss, and finally by Werner Sämann as Chairman and CEO. The company did not have its own Managing Director, but had separate MDs for Land Rover Vehicles and Land Rover UK sales. Peter Kinnaird became the Managing Director for Land Rover's UK sales company and was succeeded in 1999 by Mike Wright. Between 1994 and 1999, Ian Robertson was MD of Land Rover Vehicles; he was succeeded by Dieter Udelhoven.

The situation was different under Ford, with an extra layer of management being introduced because Land Rover belonged to Ford's Premier Automotive Group. This had its own Chief Executive Officer, initially Wolfgang Reitzle (who had left BMW for Ford), then Mark Fields and subsequently Lewis Booth.

Jaguar Land Rover also had its own Chairman under Ford. The first one was Bob Dover, a former Land Rover Director of Manufacturing. Joe Greenwell took over in 2003, then Geoff Polites had the job from September 2005, now with the title

Tony Gilroy did sterling work as Land Rover's Managing Director in the 1980s.

Phil Popham became Land Rover Managing Director in 2006, and subsequently Director of JLR Group Sales Operations.

of Chief Executive Officer for Jaguar Land Rover. He retired through ill health in 2008 and David Smith took over, steering JLR through the Tata takeover. Smith stood down in early 2010 and his replacement was Dr Ralf Speth, who had been Vice-President of Land Rover during the BMW period.

In 2003, a new post of Land Rover Managing Director was created. Matthew Taylor was the first appointee, handing over the reins to Phil Popham in 2006. Reorganization under Tata ownership saw the individual National Sales Companies given their own Managing Directors; in the UK, the job went initially to John Edwards, in 2010 to Colin Green and in 2011 to Jeremy Hicks.

The Men in Charge: Engineering and Design

Traditionally, Land Rover had been run by its engineers, but by the time of the One Ten there had been some important changes. Heavily influenced by the example of Ford, Land Rover had established a new Product Planning division in 1978.

This had the responsibility of collating information about customer requirements and engineering developments outside the company, identifying market trends and opportunities, and proposing new products for the company to manufacture. The new division reported directly to top management, but in effect its function was to tell the designers and engineers what they would have to come up with next. Many engineers saw this as unwarranted interference and Product Planning was not universally liked in its early years.

Within Product Planning came another new group of people, united under the title of Brand Management. This was a group whose job it was to ensure that proposed changes to the products would not compromise their image or usefulness and so harm sales. Once again, Brand Management was not universally popular in the beginning, but by the time of the Defender in 1990 it was guiding the important move towards leisure-market derivatives.

Meanwhile, the Chief Engineer remained the man in charge of engineering and associated development. Though his title changed more than once during the coil-sprung era, and his precise sphere of influence was also redefined more than once, he remained the key figure.

During the development phase of the Stage 2 models in the late 1970s, Land Rover's Chief Engineer had been Tom Barton. Barton had worked in a very junior capacity on the development of the original Land Rover in 1948, had gained a huge wealth of invaluable experience of the brand over the years, and had risen through the ranks until he reached the top job. But he had initially been opposed to the use of coil springs on Land Rovers and, perhaps reluctantly, allowed himself to be converted when it became obvious that this was the way forwards.

His successor in 1981 was Mike Broadhead, who had been his assistant and took over as Chief Engineer when Barton retired. Broadhead had already had a major influence on the design of the coil-sprung range, in particular leaving his mark on the short-wheelbase derivative, the Ninety. He oversaw the launch of the One Ten in 1983 and of the Ninety a year later, but he did not get on with Land Rover's then Managing Director, Tony Gilroy, so chose to leave the company in 1984.

Mike Broadhead's successor, with the title of Director of Product Engineering, was Bill Morris, another long-serving Land Rover engineer who oversaw the move into turbocharged diesel engines during the 1980s. However, the change to a merged Rover Cars and Land Rover under British Aerospace also brought another layer of management and Stan Manton was put in as overall Engineering Director for the Rover Group. As the group was restructured, Manton left and Steve Schlemmer briefly took over as head of Land Rover vehicle engineering. By summer 1990, Alex Stephenson had come in from Perkins Diesels to assume the post of overall Director of Engineering.

When BMW took over in 1994, it put Burkhard Goeschel in as Land Rover Chief Engineer. Under Ford, the top engineering job at Land Rover went to Steve Ross, an American who had overseen that company's very successful Explorer Sport Utility Vehicle (SUV) programme. Then, when Tata took over in 2008, Land Rover brought in Murray Dietsch, a former Ford employee who, although Australian by birth, had been strongly influential within Land Rover's South African operation.

The styling of the original One Ten was overseen by David Bache, the former Rover styling chief who had become British Leyland's Director of Corporate Styling Services. However, more directly responsible for the One Ten styling was Tony Poole, who remained with the company after Bache left in 1981 following a row with Harold Musgrove, then Director of Manufacturing and later Chairman of the Austin Rover Group.

From 1985, Land Rover styling took place under the supervision of Dave Evans, who had joined the company from Chrysler UK when that company had closed down. He remained in charge of styling – by then called Design – under BMW. When Ford took over, the top job went to Geoff Upex; from 2007 he handed over the reins to Gerry McGovern, who had worked on the original Freelander. However, the man with direct responsibility for Defender in this period was Studio Director Dave Saddington.

ABOVE: **Bill Morris was in charge of Land Rover engineering in the late 1980s.**

FAR LEFT: **Murray Dietsch took charge of Land Rover engineering under Tata.**

LEFT: **Dave Saddington became Studio Director for Defender in the Ford era.**

The Manufacturing Plants

When the One Ten entered production, Land Rover's headquarters was at the old Rover Company plant in Lode Lane, Solihull, in the West Midlands. Rover car production had ended there in the last months of December 1981, being transferred to the old Morris plant at Cowley; meanwhile, production of the Triumph sports cars built there from 1979 had also come to an end. Solihull was, and remains today, completely dedicated to Land Rover manufacture.

At the start of the One Ten era, Solihull was primarily an assembly site, depending on a number of 'feeder' or 'satellite'

LEFT: **Defenders on the assembly line at Solihull in 1990.**

BELOW LEFT: **This was Land Rover's V8 engine assembly line at the end of the 1980s.**

BELOW: **The assembly line at Solihull has always depended heavily on people rather than machines. This was it in the early 1980s.**

factories in the surrounding area that manufactured subassemblies. So diesel engines and V8 petrol engines were brought in from Acocks Green, while 4-cylinder petrol engines came from Tyseley. Transmissions came from Percy Road, axles from Perry Bar, suspension and bell-housings and gearbox casings from Tyburn Road in Erdington. Clay Lane in Coventry provided cabs and hardtops. All these items reached Solihull by road because the site did not have – and still does not have – its own rail terminal.

All the 'satellite' factories were progressively closed in 1984–5, with their functions redistributed around the Solihull site. Outside suppliers continued to provide some major

Still dependent on people, these were the Defender production lines in 2007.

items, however, and notable among them was John Thompson Pressings in Wolverhampton, which manufactured all the chassis frames. From 2006, Defender engine supply arrangements changed, as Ford diesel engines began to arrive from the Ford plant at Dagenham.

The Overseas Assembly Plants

A proportion of coil-sprung Land Rovers has always been assembled abroad, usually to avoid prohibitive import duties on fully built vehicles. The figures have varied considerably over the years, but during the period of the coil-sprung Land Rovers have usually been between 3,000 and 5,000. Coil-sprung models were initially exported in knock-down kit form through Export Packing in Banbury, but from the mid-1980s the operation was taken back in-house and based at Solihull. The number of overseas assembly operations also dwindled from the mid-1980s, mainly as a result of a changing political climate, although global complete knock-down (CKD) figures were on the rise during the 1990s.

These overseas assembly operations typically added a proportion of locally manufactured content to the material sent out from Solihull, although the proportions varied from one plant to another. Two overseas assembly plants, in Australia and South Africa, depended heavily on locally manufactured items and were able during the coil-sprung Land Rover period to develop distinctive models of their own.

The major overseas assembly operations were in Australia, Brazil, South Africa and Turkey. The Australian operation was based at Moorebank in the suburbs of Sydney and was run by Jaguar Rover Australia (JRA). Though relatively new in this location, it continued a tradition of building Land Rovers in Australia that stretched back to the start of the 1950s. One Ten assembly began there in the third quarter of 1984 and continued right through until JRA ran into difficulties in 1992, by which time the models were Defenders. The Australian operation also developed its own vehicles – notably the 120in chassis and the special Perentie 4×4 and 6×6 military vehicles. Land Rover sales in Australia after 1984 mixed locally assembled and completely built-up (CBU) imports from Britain up to 1992, as well as South African-assembled models from 2003. Between 1996 and 1998, a further batch of military models was also assembled at the British Aerospace plant in Adelaide.

The Brazilian operation opened in 1998 at the Karmann plant in São Paulo, assembling primarily civilian models for the local market. When Ford bought Land Rover from BMW in 2000, Land Rover do Brasil merged with the Brazilian Ford operation, but CKD assembly remained at São Paulo. The Brazilian plant handled a large military contract for the Brazilian armed forces in 2003–4 and was much respected for the quality of its products. By mid-2004, it was assembling around 1,500 Defenders every year, while some 500 units of Land Rover's other models were imported as fully built vehicles from Solihull. However, the Brazilian assembly operation was closed at the end of 2005 as part of a rationalization scheme under Ford.

The South African assembly operation was also closed in 2005 under the same scheme. Land Rovers had been assembled in South Africa since October 1963. Assembly of

Defenders on the assembly line in South Africa at the Ford-owned assembly plant in Silverton.

The Otokar CKD operation in Turkey was established on the back of military and other Government requirements, but it also built civilian models. This is a One Ten Station Wagon, distinguishable from the Solihull product in this view only by the small Otokar decals behind the front wheels.

coil-sprung models was initially at Blackheath in Cape Province, but was moved to the BMW plant at Rosslyn, Pretoria, in 1998, then moved again to the Ford-owned assembly plant at Silverton in mid-2001. It assembled about 1,500 Defenders a year. Like the Australian operation, the South African one developed a number of its own special models. Notable among them were the Defender 130 Game Viewer and the Defender 147.

The Turkish CKD operation was carried out by Otokar, initially in Istanbul and later in Arifeye. One Ten assembly began in 1987 and was primarily for the military and for Government agencies in Turkey. Otokar developed a number of its own special body variants and by the turn of the new century, Defender build volumes were around 2,000 a year.

There were several smaller operations. In Jordan, a CKD operation was established at Ma'an in the south of the country in mid-2005, as a joint venture with the Olé Automotive Trading Company of Amman. Build volumes were claimed to be as high as 2,000 a year, but this was probably a short-term operation to deal with a specific military contract. The initial plan was to build 5,000 vehicles, all Defender 110s, and these seem to have included pick-up, Station Wagon and window hardtop variants.

In Kenya, the Cooper Motor Corporation of Nairobi oversaw a small-volume operation assembling around thirty-five Defenders a year. This was actually carried out by Kenya Vehicle Manufacturing, which also assembled Mazda and Nissan trucks, and its main function was to deliver special derivatives for local use. The best known was the 135in chassis intended for safari tour buses. There was a slightly larger operation in Malaysia, where 200 to 300 Defenders were assembled each year by the Ford-owned Associated Motor Industries in Shaha Alam, Selangor; the operation later moved to the Inokom plant in Kulim.

Considerably larger was the CKD operation in Pakistan, which began in 2003 and assembled around 1,500 Defenders a year. Assembly was at the Ghandara Nissan plant in Karachi and the vehicles were for both civilian and military use. Assembly of Defenders in Thailand began in 2011. Then there was another low-volume operation, building around twenty to twenty-five Defenders a year, in Zimbabwe. Land Rover's CKD partner was latterly the Quest Motor Corporation, which took over the operation from Leyland in 1993.

Special Vehicles

In the very earliest days of the One Ten, Land Rover still operated its Approval system for conversions and adaptations of the vehicle. This had been established as long ago as 1957 – the department responsible for 'approving' conversions was the Special Projects Department. Its function was to act as a central point where aftermarket specialists who wanted to market a conversion or adaptation for the Land Rover could get their work tested. If it did not compromise the integrity of the vehicle, it would be awarded Land Rover Approval, which meant that Land Rover would honour their standard warranty on a conversion. They would also help to promote its availability through their showrooms and by displaying demonstrator vehicles at the 'field days' held regularly in most countries to attract customers.

However, the founder of the Special Projects Department, George Mackie, retired in 1981 and the department seemed to drift for a while. It was probably late in 1984 that a member of the Product Planning Department, Roland Maturi, put together a plan for reorganizing Special Projects. He presented it to Land Rover MD Tony Gilroy, who liked Maturi's vision and the additional profits it promised. Maturi was duly appointed to head the reorganized department, which from July 1985 was formally known as Special Vehicle Operations.

Special Vehicle Operations took over the functions of the department it replaced, but to them it added the development of its own conversions. Its longer-term aim, which became clearer towards the end of the 1980s, was to bring as much as possible of the conversions business in-house. Though this would create greater profits for Land Rover, it did not please some of the converters, who saw their market slipping away from them.

SVO's new product range was spearheaded in April 1985 by the backhoe conversion, made in conjunction with Polymark Beaver Ltd. However, the main focus of the new division's work was on a modular series of box bodies marketed under the name of Quadtec and designed to exploit the possibilities of the One Two Seven range. These were built in-house and could be fitted with all kinds of specialist equipment, some of which was of course sourced from outside the company. Quadtec 1 was a low-roof box with a crew-cab chassis; Quadtec 2 was a high-roof box with a crew-cab

This was one of SVO's first products, a Quadtec 2 body on the One Two Seven Crew Cab.

chassis; Quadtec 3 was a low-roof box with a standard cab; and Quadtec 4 was a high-roof box with a standard cab. Doors could be fitted more or less anywhere the buyer wanted.

Roland Maturi's declared plan was to make Special Vehicles into Land Rover's 'fourth marque' alongside the Defender, Discovery and Range Rover, but he left the company in 1992. SVO was reorganized yet again, this time as Land Rover Special Vehicles (LRSV). It added the old Special Installations Division (responsible for such things as winch installation) to its empire, then expanded further to embrace another new area of special conversions. This became apparent during 1993 when Land Rover announced the Autobiography custom-finishing scheme for Range Rovers, but in fact LRSV had also built the slightly earlier SV90 special-edition Defender.

Special Vehicles (as it is usually called) went on to take responsibility for all the UK-originated volume-build special derivatives of Defender, adding the special features to part-finished vehicles taken from the assembly lines. The North American Specification models (see Chapters 6 and 7) were also completed by Special Vehicles. The division was responsible for equipping those Defenders used as support vehicles on the Camel Trophy and G4 Challenge adventure events. It has continued to act as a focal point for aftermarket conversions, advising and assisting specialist converters where appropriate, and coordinating manufacture in cases where several specialists have made individual contributions to a single vehicle. However, the volume of commercial conversions has diminished over the years, largely because of the cost associated with testing and individual vehicle approval.

A Special Vehicles commercial vehicle normally carries identifying badges on the front wings. Initially, these were decals, but from 1998 they were three-dimensional aluminium items. All Defenders built by Special Vehicles, including the special-edition models, have an identifying plate under the bonnet with the contract number and date of build.

There was yet another reorganization in March 2012, when LRSV expanded to take on custom-building duties for Jaguar Cars as well, which of course had been Land Rover's close partner ever since the Ford takeover had brought them under common ownership in 2000. Under Tata ownership after 2008, when the company had been renamed Jaguar Land Rover, this closeness was further emphasized. The new name for the division was ETO – Engineered To Order.

LRSV had its own distinctive logo, which it applied to the Defenders with which it had been involved.

PRODUCTION FIGURES

Land Rover quotes its production figures by calendar year, not by model year, so it is impossible to separate out individual model year build figures. The figures for 1983–5 also include a proportion of Series III models, which were on their production run-out. These figures were the best available at the time of publication, but should not be considered definitive: Land Rover has a habit of recalculating its production figures from time to time.

1983	30,093	**1991**	18,915	**1999**	24,589	**2007**	22,328[2]
1984	25,604	**1992**	17,626	**2000**	28,798	**2008**	22,369
1985	31,046	**1993**	17,482	**2001**	24,351	**2009**	13,770
1986	20,957	**1994**	22,265	**2002**	25,868	**2010**	15,499[3]
1987	19,933	**1995**	27,598	**2003**	21,193	**2011**	19,768
1988	21,928	**1996**	31,724	**2004**	22,151	**2012**	n/k
1989	23,437	**1997**	36,169	**2005**	18,190		
1990	21,210	**1998**	31,710	**2006**	19,294[1]		

[1] Plus 4,824 KD. [2] Plus 1,200 KD. [3] Plus 1,740 KD.

CHAPTER TWO

GETTING THERE: THE HYBRID PROTOTYPES

The critical factor in the further development of the Land Rover was actually the huge success of the Range Rover, which was introduced in 1970. This had three major technological advances over the contemporary Land Rovers. First, it had permanent four-wheel drive, which not only provided it with the traditional Land Rover rough-terrain ability, but also improved handling on the road. It also did away with the need for heavy-duty axles with powerful engines, so reducing unsprung weight and permitting a more car-like ride.

The second of the Range Rover's major advances was vastly improved stopping power, brought about by an all-disc braking system in place of the all-drum system still used on Land Rovers. The third was its reliance on coil-spring suspension, which gave a comfortable ride both on and off the road.

At Solihull, the potential advantages for future Land Rovers of permanent four-wheel drive and disc brakes were obvious. The sticking point, however, was coil springs. Land Rover's somewhat autocratic Technical Director, Tom Barton, was implacably opposed to coil springs on future models. His view was that the rough ride given by leaf springs inhibited the driver from driving too fast over rough terrain and so damaging the vehicle. Coil springs, he was convinced, would give a ride that was too comfortable, so putting the vehicle at risk from a driver who lacked mechanical sympathy.

In the normal course of events, some preliminary work on an eventual successor follows almost immediately on the launch of a new model, partly so that the good ideas that went unused for that vehicle do not get lost. So it was with the Land Rover, with the result that after the Series III models had entered production in autumn 1971, some work was done on a possible Series IV model. The fact that it would have leaf-sprung suspension seems not to have been questioned.

Series IV looked at a number of other new ideas, though, and by 1974 had evolved into a new project that attracted the code name of SD5. This reflected the fact that Land Rover now belonged to the Specialist Division of British Leyland and it was that division's technical chief, Spen King, who allocated the code name; it was simply the fifth new vehicle project since the Specialist Division had been established in 1971.

By this stage, the Range Rover had demonstrated beyond reasonable doubt that the additional comfort of coil-spring suspension did not lead to excessive damage in off-road driving. It was also true that Spen King was the man who had originally proposed using coil springs for the Range Rover and it may be that he encouraged the Land Rover engineers to look again at their use on future utility models. Tom Barton, perhaps sensing that he was now fighting a losing battle, asked his deputy Mike Broadhead to oversee the SD5 project – and coil springs found their way into the specification.

They were only used on the front wheels of the 'mule' prototypes that were built by converting Series IIIs, perhaps because leaf springs seemed like a cheaper option for the rear. The issue was that the Range Rover incorporated a self-levelling strut which pumped itself up to keep the vehicle level when there was a heavy load in the back. Land Rovers, designed to carry such loads all the time, would certainly need one – and the additional expense and complication may have been seen as undesirable.

However, the whole way in which Land Rover approached new products was changing in the early 1970s. Traditionally, ideas for new products had come from the engineers or the management, perhaps, after discussion, being built into prototypes for testing. If they worked, they would then become part of a formal engineering project, which might eventually result in a new product. However, one way in which British Leyland was trying to streamline its operations at this time was by borrowing ideas from Ford, which led the market in Britain. Ford had for many years had a department that was

The raw material: the Series III and the Range Rover were Land Rover's products in the mid-1970s, and the new Land Rover combined elements of the two vehicles.

dedicated to examining future product opportunities, so at Land Rover a similar department was established with the name of Product Planning. At its head was Roland Maturi, a former Ford employee.

So the SD5 project, promising as it was, was halted in 1975 as Product Planning turned their attention to future Land Rovers. A major document – the Assumptions Document – was drawn up within Product Planning, with the idea that this should be the basis of future work on new Land Rovers. Many of the ideas that went into this came out of a world survey of Land Rover's markets that had been done between 1972 and 1974. Unfortunately, no copy of this vital Assumptions Document has yet come to light. However, Roger Crathorne, who worked on it during a spell in the Product Planning department, remembers that it envisaged Land Rover product development taking place in two stages.

Despite calls for a larger diesel engine from several markets around the world, the focus of Stage 1 was on replacing the old 6-cylinder petrol engine. Land Rover had just been taken off the Arab embargo list and was keen to get back into the Gulf markets, so this became the top priority. Stage 1 envisaged uprating the existing Series III models by fitting the 3.5-litre V8 engine from the Range Rover, together with its gearbox and permanent four-wheel-drive system. Stage 2 envisaged a further upgrade, this time by fitting coil springs to what was in effect the existing Land Rover. As there was very limited funding available, Mike Broadhead endorsed the 'minimum change' approach that the Product Planners were recommending.

Work began quite quickly on the Stage 1 proposal, which actually retained that name when it entered production in 1979. Stage 2 was of course further in the future and it was not until 1976 that work began on that. The change to coil springs was going to need considerable additional development, so Mike Broadhead asked Bill Morris, then Chief Engineer in charge of the product development teams, to take the project on.

In the meantime, Roger Crathorne had moved out of Product Planning and back into the Vehicle Development side of the Engineering Department, where he was working on a variety of trouble-shooting projects. It was during 1975 that he had idly used up some spare time at his drawing board by sketching up a coil-sprung Land Rover. In effect, he had simply amalgamated the relevant parts of the General Arrangement drawings of a long-wheelbase Land Rover and a Range Rover. The idea seemed logical to him, as he knew the benefits of coil springs from working on the original Range Rover project in the late 1960s. However, Technical Director Tom Barton caught sight of what he had done and was very definitely not impressed. Roger was given a disciplinary interview and was told not to waste company time by drawing up a vehicle that was not part of company policy.

So when he was asked to take the idea of a coil-sprung Land Rover further, Bill Morris remembered this incident, and suggested that Roger should be given the job of building the 'mule' prototypes that would have to be the first stage of the new project. By the autumn of 1976, Roger had been given the job of building four prototypes. They had to be completed quite quickly, because they would be needed in early 1977 to demonstrate to the British Leyland (BL) Board what Land Rover had in mind for the future. Only after that, and only if they liked what they saw, would the BL Board

ABOVE: **The Stage 1 V8 was the first stage in upgrading the basic Land Rover.**

RIGHT: **The Range Rover chassis was the starting point for the new coil-sprung Land Rovers.**

approve the funding for a full development programme of the Stage 2 vehicle.

Driven by Product Planning's Assumptions Document, the plan was for two different sizes of Land Rover, as had been the case ever since 1953. But the short-wheelbase models, then with an 88in (2,235mm) wheelbase, were generally seen as too small. Tom Barton in particular favoured an intermediate wheelbase of 100in (2,540mm) – the same size as the Range Rover's – with a long-wheelbase model broadly similar to the existing 109in (2,769mm) of the Series III Land Rover. As the Range Rover chassis had already been stretched to 110in (2,794mm) for some special applications, 110in (2,794mm) became the long-wheelbase size. However, the idea of a short-wheelbase model had not been completely excluded, and the wheelbase dimension for this was set somewhat arbitrarily at 90in (2,286mm) – very slightly larger than the 88in that was considered too small.

As a result, the four 'mules', or hybrid prototypes, were to consist of two with a 100in wheelbase because this was the major new development, one with a 110in wheelbase and one with a 90in wheelbase. Of the 100in models, one had to be a Station Wagon, to see whether the new dimension would give enough passenger space and comfort.

As Roger remembered it for the July 2003 issue of *LAND ROVER enthusiast* magazine:

> *Off I went to see Maurice Hanson, who was our Production Director, and I asked him for four Range Rover rolling chassis and four body 'kits' for Land Rovers. I told him what I was doing and Maurice said, 'It won't fit!' but he let me have what I wanted.*
>
> *I came in on a Saturday morning and I remember watching a fork-lift bringing four sets of painted Land Rover panels into the Experimental Shop. They were just as they'd come out of the Paint Shop, on their trollies, and they were all grey because they'd been doing a run of grey vehicles that week!*

All the work was done in the Experimental Department and Roger had a team of fitters to carry out his instructions. As a first stage, the Land Rover bodies were assembled; a photograph taken on 14 October 1976 shows one of these resting on the chassis of a Range Rover, to give a clearer idea of where the necessary cutting and shutting would have to take place. Among other things, it was already clear that the tyres would project beyond the body sides because the Range Rover axles were wider than the Land Rover ones for which the body had been designed. In keeping with the philosophy of 'minimum change', the solution was not to cut the body down the middle and widen it, but rather to fit wheel-arch extensions to cover the tyres. The job of drawing these up was given to Tony Poole, then in charge of Land Rover work in the Styling Department. Surviving photographs from late November 1976 show both simple and more elaborate proposals, in each case using aluminium alloy.

The 90in (2,286mm) hybrid was built up as a soft-top, while the 110in (2,794mm) model was a pick-up-bodied truck cab. Both had standard Range Rover V8 power trains. Of the two

The basic principle was to fit a Land Rover body to a Range Rover chassis. The Land Rover body was designed for a 109in (2,769mm) wheelbase and the Range Rover chassis had a 100in (2,540mm) wheelbase, so the body had to be cut down. This is one of the four 'hybrid' prototypes at the very early build stage: the Series III body has been lowered onto the Range Rover chassis and the next stage is to mark up where sections will be removed. The picture was taken on 14 October 1976 and the vehicle probably became the 100in soft-top.

100in (2,540mm) vehicles, one was built up as a four-door Station Wagon, although it had a canvas roof which saved the time and cost of modifying standard production panels. This also had a Range Rover V8 power train. The second 100in was built as a two-door soft-top, but this had an experimental power train with a Land Rover 2.25-litre petrol engine and remotely mounted gearbox, at Mike Broadhead's request. All four hybrids had right-hand drive and all four had single-line braking systems plumbed into the Range Rover's twin-piston front calipers; more details of the individual vehicles are shown below.

Once the four vehicles had been completed, they were put through a brief test programme to iron out minor problems. The demonstration for the BL Board was then set up at Eastnor Castle in Herefordshire, where Land Rover had been testing vehicles off-road since 1961. Two vehicles were selected as the primary demonstrators, and these were the 100in soft-top Station Wagon and the 110in truck cab. Their job was to demonstrate to the Board that the proposed new Land Rovers were superior in their abilities to existing production models.

It was an important exercise that lasted for two or three weeks. The Board members did not arrive en masse, but singly or in pairs as their schedules permitted. Tom Barton stayed at Eastnor for the whole period in his caravan, which he had towed down behind the hybrid 90. As the engineers had known they would, the hybrids proved superior to current production types in every way. 'There was one particular section we'd set up,' remembered Roger Crathorne, 'where the leaf-sprung 109 lifted its wheels and got stuck, but the 110 just ploughed on through.' The board approved, and 'they actually did the sign-off in the snooker room at Eastnor Castle.'

The Four Hybrids

None of these hybrids was originally registered for the road and all were running on trade plates at the time of the demonstration to the BL Board in early 1977. Unusually, none of them had any identifying prototype numbers either and they have never been traced in any surviving company records. All of them had a grille panel that was flush with the wing fronts, as had already been designed for the Stage I V8 models, but the grille itself was a Series III type on some and a Stage I V8 type on others.

The 90

Although Roger Crathorne remembers building the 90 as a right-hand drive vehicle, no pictures survive of it in this condition. Pictures dated 19 January 1977 show what is very clearly

Nobody is quite sure, but this is probably the original 90in (2,286mm) hybrid after being kitted out as a military demonstrator. If it is that vehicle, then it had also been converted to left-hand drive and fitted with a Series III grille by the time of this picture.

This is the 90in hybrid, still retaining the staggered rear damper arrangement of the Range Rover chassis (*top*), together with its self-levelling strut and all-disc braking system (*above*). There have clearly been problems with the wheel-arch extension already!

This is the 90in again. The fan cowl is rather crude, but everything else seems to have fitted rather nicely around the Range Rover V8 engine.

a left-hand-drive 90, kitted out as a military vehicle. At this early date, it can only be the original hybrid 90.

Several people have dim recollections that Tom Barton offered a short-wheelbase coil-sprung model to meet a French military requirement very early in the Stage 2 story, almost before any vehicles were available. So it seems probable that Tom had the hybrid 90 converted to left-hand drive and militarized so that he could show something to the French as early as possible. Although no documentary or photographic evidence has so far confirmed this, it is certain that there was a 'French prototype' by April 1977, when it was mentioned in the Stage 2 Project Notes, which also record that more French vehicles were to be built. As the next chapter shows, these were in fact built on a 100in wheelbase.

After the BL Board demonstrations were over, the hybrid 90 may have remained at Eastnor Castle. Pictures from April 1977 show what is almost certainly this vehicle back at Solihull, with light damage after off-road use. During 1978, it was used for various assessment exercises and also to demonstrate what off-road vehicles were all about to new engineers who joined Land Rover at the time.

However, during this period, the idea of a 90in Land Rover fell right out of favour and, as the next chapter shows, the engineers devoted their attentions to the 100in and 110in variants of the Stage 2 models. Roger Crathorne remembers Tom Barton stating very firmly some time in 1978 that 'Land Rover will never build a 90in short wheelbase', and giving orders for the vehicle to be scrapped.

The 100in Station Wagon

This hybrid was quite a remarkable vehicle and was the only one of the four not to have a standard Land Rover windscreen. As Roger Crathorne explained in 2003:

> *We had to meet some European regulations about being able to see overslung traffic lights, and you just couldn't see them through the little letterbox of our split windscreen. Our Swiss importers had a windscreen made to show us what was needed, with more height at the top, and they sent it across. So we fitted it to this vehicle.*

With the 100in wheelbase, the rear overhang was considerably reduced and so the rear-mounted fuel tank of the Range Rover chassis was removed and a pair of Land Rover underseat tanks were fitted. To save time and costs, only the lower panels of the rear body were used and the upper

This second 100in hybrid model retained its V8 engine, but was configured as a Station Wagon. It was quicker and cheaper to have a canvas roof made than to modify a 109in roof, so it was completed that way. Roger Crathrone was pictured with the vehicle more than twenty years after he had overseen its construction.

section and the whole of the roof were discarded. Wooden battens were made up to get everything to fit together where the body had been shortened behind the rear wheel arches. To make the vehicle weatherproof, the simplest and quickest solution was to have a special canvas tilt made up, with the result that this hybrid looked like no other Land Rover of the time.

To restore the body strength that a fixed roof would have provided, a German-specification cross-brace was fitted behind the front seats. However, there was insufficient room for a standard Station Wagon bench seat ahead of the rear axle, which of course was in a different position from the axle of a 109. Enough room was available for a pair of individual striped bucket seats from an MGB sports car – and so they were matched by a similar pair of front seats. Together with a side-hinged tailgate (a rarely seen Land Rover military option of the time), the resulting impression was of a vehicle rather different from a standard Land Rover, which helped to create the right impression on the members of the BL Board. There was, too, some thinking going on about a more sporty Land Rover at the time and this dovetailed neatly into that.

As built, the hybrid 100in Station Wagon had aluminium wheel-arch eyebrows at the rear, but 'I took them off the first time we took it round Eastnor,' remembered Roger Crathorne in 2003. 'They very quickly caught in deep ruts and started to tear away.' The Eastnor experience also prompted another modification: there was more axle articulation than the rear coil springs could handle, and they tended to drop out of their upper mountings. So Roger fixed that, by putting a circlip around the topmost coil on each side to hold it to the chassis bracket. At the time of writing, thirty-five years later, those rusty circlips were still there.

After its successful demonstration to the BL Board, this vehicle was used very heavily at Eastnor as an off-road hack and was kept there for many years. When it was eventually pensioned off, Roger Crathorne managed to prevent it from being scrapped, steering it into the hands of the Dunsfold Collection in Surrey. It was still there in 2012.

The 100in Soft-top

The soft-top 100in was completed a little later than the other three hybrids because of its special power-train arrangements. Where the others all had standard Range Rover V8 engines and 4-speed LT95 gearboxes, this one had a 2.25-litre petrol engine and a remotely mounted transmission. It was not completed in time for the Eastnor Castle demonstrations to the BL Board.

The 2.25-litre petrol engine was of course standard Land Rover fare of the time and was therefore not exceptional. However, the remotely mounted gearbox, with a propshaft between the engine and the main gearbox, was interesting. The idea behind it was mainly to improve refinement. More than a decade and a half earlier, Mike Broadhead had experimented with a remotely mounted transmission in one of the big 129in (3,277mm) Land Rover prototypes. However, in that case, the main gearbox had been mounted conventionally

This is the 100in soft-top prototype. The join in the modified rear side panel is just visible.

Many years later, the 100in soft-top had become a 90in soft-top. It had been fitted from the start with a 4-cylinder petrol engine.

behind the engine and there had been a short propshaft between main and transfer gearboxes.

When asked in 2009, Mike simply could not remember anything about it. His best guess was that it had been cooked up as a Tom Barton special that was pushed through the workshop 'without involving the cumbersome "system", which bogged the rest of us down!'

The vehicle probably outlived its usefulness as an engineering prototype quite quickly, but what happened to it next is very far from clear. A best guess is that the engineers had a sudden need for a 90 prototype during 1978, perhaps because thinking was now returning to the idea of 90 and 110in wheelbases rather than 100in and 110in sizes. However, the original hybrid 90 had already been scrapped. So it looks as if the 100in soft-top was rebuilt on a 90in wheelbase as a temporary expedient. The twin underseat fuel tanks were also switched for a rear-mounted Range Rover tank.

The next stage in the story involves Dennis Roberts, who was then Finance manager of the Engineering Department. Dennis had a second home in Somerset and opposite the house was a cider orchard on a steep slope. When the orchard was put up for sale in 1976, Dennis felt he had little option but to buy it if he wanted to preserve the pleasant view from his house.

> *At that time, cider was still being made locally and the apples from this land were regularly used. Our problem was that at the time of purchase these apples were ready to collect and we had no means of transporting them from the top of the hill down to road level.*
>
> *The obvious solution was a Land Rover, so I asked Tom Barton ... if we had a fleet vehicle I could borrow for the weekend. His reaction was to suggest that as all the work would be off-road, I could use the 100in unregistered concept vehicle. It was brought to Somerset on trade plates and used for several weekends to do the job in hand, spending the intermediate time locked securely in our garage. Soon after this ... I was approached to rejoin Ford and so took my leave of Rover with, alas, the concept vehicle still safely locked away in Somerset.*

The Range Rover chassis was extended by 10in (254mm) to make a 110in prototype. The bolt-on 'mudguard' was one idea for extending the body to cover the wider tracks of the Range Rover chassis. This vehicle still survives in the Dunsfold Collection, but is now a hardtop.

There it stayed for two or three years, but Dennis wanted to legitimize the situation, so he contacted Tom Barton again. He offered to return it to Solihull, but also pointed out that he would be happy to buy it because it suited his continuing need for a Land Rover. Tom suggested he should make an offer of about £250 for it, which he did. This was accepted, 'and confirmed by Tom phoning to tell me the bad news was I would have to pay the VAT!' The additional VAT came to £37.50 and the delivery note from Solihull to Dennis still survives. On it, the vehicle is described as a ' "Lash-Up" Land Rover chassis'!

That was in March 1980 and Dennis decided to make it road-legal at the same time. As the original wheel-arch eyebrows had long since disappeared, Dennis had some new ones made up out of rubber and on 1 July 1980 he had the vehicle registered in Somerset as RYB 654V. By that stage, it had acquired a chassis plate reading '90/1'.

This vehicle still survived at the time of writing, now in the hands of Dennis' son-in-law.

The 110in

The extra 10in (254mm) of wheelbase were inserted into the centre of the chassis, in the same way as was already being done when 110in Range Rover chassis were being created for ambulances. The Range Rover fuel tank was left in its standard position at the rear.

This vehicle was built as a truck cab with pick-up body and initially had large alloy 'blisters' to widen the body and cover the rear wheels. It was the second of the hybrids that demonstrated the coil-sprung Land Rovers to the BL Board in early 1977. Subsequently it became an off-road hack at Eastnor Castle.

During its time at Eastnor, it was converted to window hardtop configuration, with filler panels at the front of the roof and behind the doors to make up for the inch of difference between the 109in wheelbase that the panels were intended for and the 110in wheelbase they were on. It was also given a one-piece windscreen, made at Solihull but following the pattern of the Swiss prototype used on the hybrid 100in Station Wagon.

This fourth hybrid later joined the 100in Station Wagon in the Dunsfold Collection in Surrey, still surviving at the time of writing.

CHAPTER THREE

DESIGN AND DEVELOPMENT, 1977–1983

The BL Board's approval of the Stage 2 hybrids at those demonstrations in early 1977 was a source of great relief to management and engineers alike at Solihull. Not only the Stage 2 Land Rover, but also plans for the future of the Range Rover, could now rely on funding which was in fact being provided by the Government as an investment in Land Rover's long-term future.

Plans were already in hand for the company to become a standalone business unit within British Leyland and Land Rover Ltd was established in July 1978 with its own Board and Managing Director. In August, the details of the Government investment plan were announced. The first £30 million would become available immediately and would cover the launch of the Stage 1 V8 109 models, plus increases in production capacity for both Land Rovers and Range Rovers. The second £250 million was earmarked for later. It would, in fact, cover the cost of getting the Stage 2 models into production, plus some additional development on the Range Rover. However, nothing became public at this time about what Land Rover had in mind for Stage 2; there was no value in alerting competitors to the company's future product plans.

However, the introduction of the new coil-sprung Land Rovers was still a very long way off – the design and development phase would last for the best part of six years. Though this was not out of alignment with industry norms at the time, it was very long in comparison with the design and development phase of the original Land Rover (about six months in 1947–8).

Work on the Stage 2 Land Rovers formally began at Solihull in April 1977, more or less as soon as humanly possible after the BL Board had approved the funding. The key date seems to have been 27 April, when a project meeting was held to review what had to be done. By that time, the 90in (2,286mm) had disappeared completely from the programme, which now focused on 100in (2,540mm) and 110in (2,794mm) wheelbases. There was to be maximum commonality between the two.

It was already clear that the chassis side members would have to be retooled because the Range Rover chassis as it stood was not strong enough, and there was also some hope that the new 110in chassis might also be used for the four-door Range Rover when it entered production. (In practice, the redesigned side members were too deep for the Range Rover and the four-door model went ahead with the existing 100in chassis.)

As for engines, there were to be three. One was the 3.5-litre V8, one was a 4-cylinder petrol – described as a 2.4-litre but certainly the one that entered production as the 2.5-litre type – and a diesel, which was mysteriously described as a 2.7-litre. Land Rover was gradually changing over from imperial measurements to metric ones and it was already clear that the engines would be fully metricated. An issue was to decide on a metrication policy for the rest of the vehicle.

Gearboxes would be the LT95 behind the V8 engine and the Santana-built heavy-duty LT85 5-speed with an LT230 transfer gearbox (which was still under development at Solihull) behind the 4-cylinder engines. The V8-engined models would have disc brakes at the front, but others would have drums all round. It was likely that production of Stage 2 models would overlap with the end of Series III production, and therefore commonality between the two types had to be considered.

A project team was also assembled to develop the new vehicle, probably after that meeting in late April. Project Engineer, and therefore team leader, was Bob Lees. Working under him were Graham Fazakarley, Hugo Vernon and Denis Yates. Later, Brian Anderson also joined the team. Members of other departments were of course involved in the design of the Stage 2 Land Rovers as well. Head of the chassis section of the Drawing Office was Trevor Greenway, while Tony Poole ran the Land Rover section of the Styling Department. Specialists from other departments would have their input as required on such things as electrical systems, transmissions, engines and so on.

The First Prototypes, 1977–8

The team lost no time in building some prototypes. The first two seem to have had the 100in wheelbase and one was completed as a right-hand-drive civilian truck cab, while the other was a left-hand-drive military soft-top. Both had the traditional two-piece windscreen, which was readily available on the production lines, even though the idea of a one-piece screen was already in the air. The civilian prototype was grey and was registered NRW 682S; the military prototype was Bronze Green and became VVC 701S, with a registration date of 7 September 1977.

It looks as if no 110in prototype was built at this early stage of the project, which was almost certainly because the 100in model was seen as the more important of the two. As commonality of components was the name of the game, almost everything could be developed on the 100in chassis and work could move on to the 110in later, when only items specific to that model would need development. The 'minimum change' approach was still in place and, as overseer of the Stage 2 project, Mike Broadhead was insistent that as much as possible of the existing production Series III should be carried over.

There was plenty of design work to be done, though. During 1978, for example, the Styling Department was working on the design of the new one-piece windscreen and associated changes to the cab roof. Tony Poole remembered his frustration that the windscreen could not be extended downwards because the bulkhead ventilators were going to stay in their traditional places and have their traditional sizes. Instead, it had to be extended upwards into the roof area. This, in turn, required new solutions to meet new or forthcoming regulations in some countries about the area swept by the windscreen wipers.

Five 110in prototypes then followed in mid-1978, numbered 110-X1 to 110-X5. The first ones were completed in June and the last in August, and there was a variety of different body

ABOVE: **This was probably the very first 100in prototype, a RHD civilian-specification truck cab. Note the Series III grille, which was present probably mainly for disguise purposes. The windscreen is still a two-piece type.**

RIGHT: **VVC 701S was an early LHD 100in prototype built to a military specification with front quarter-bumpers and a window soft-top. The windscreen was still a Series III two-pane type.**

configurations along with examples powered by the 2.25-litre 4-cylinder and 3.5-litre V8 petrol engines. There may also have been diesels. These vehicles were put through the usual array of tests designed to flush out their weaknesses.

The Military 100in Prototypes, 1978–86

Meanwhile, the opportunity arose for Land Rover to tender for a very big military contract. The French Army was beginning the search for a fleet of vehicles to replace its 10,000 Hotchkiss-built Jeeps, with deliveries of the new vehicles beginning in 1981. Tom Barton had picked up on the early stages of this in early 1977, when he had arranged for the hybrid 90 to be mocked-up as a military demonstrator and had showed it to the French. Now things were looking much more interesting: trials had been scheduled and manufacturers had been invited to submit vehicles.

Bob Seager, who then ran the Military Development section at Land Rover, remembered that the company found out about the trials rather late. Undeterred, the engineers rapidly put together a 100in V8 military prototype – most likely VVC 701S, one of the first two built – and Solihull offered to join in. The trials were held near Lorient, in Brittany, and the 100 outshone its competitors, especially in the off-road sections.

The Swiss military tender called for some interesting body variants, including a four-door soft-top. This prototype was not registered until much later, when it was given an age-related number.

As the next stage of the selection process, Land Rover was invited to submit a trials batch of fifteen coil-sprung vehicles.

Preparing these vehicles was a major project in itself, especially as the Stage 2 specification was still being developed. So, to contain costs, the Stage 2 project team focused their attention on the 110in models and built no more 100in prototypes. The Military Section under Bob Seager took on responsibility for building the French trials vehicles – and of course the two development teams kept in close touch so that their findings could be shared as appropriate.

The first of the French trials vehicles was put on the road in September 1978, just a month after the last of the 110 prototypes. The rest of the batch followed over the next few months. However, the Solihull engineers were not at this stage aware of a very large fly in the ointment. Among those who were tendering for the contract was the French car maker Peugeot, who had agreed with Mercedes-Benz to build the German company's still-secret G-Wagen under licence and to fit it with their own engines. One of the Peugeot models had been among the trials vehicles at Lorient.

The Peugeot solution had great political appeal to the French, as it channelled most of the funds allocated for the new vehicles back into French companies. There was also no doubt that the G-Wagen was an excellent vehicle in its own right. Slowly it became apparent that the Peugeot P4 was the front-runner for the contract. So by early 1979, with some of the fifteen trials vehicles still unfinished, it became obvious that Land Rover was wasting its time. Bob Seager tried a last-ditch effort by fitting a Renault 2-litre petrol engine into one of the trials vehicles, reasoning that two could play at Mercedes' and Peugeot's game. Although the engine went in fairly well, Bob remembered it was 'a bit gutless'. It was not enough to tip the balance, however, and Land Rover decided to pull out of the race for the French contract.

Nevertheless, the trials fleet was not wasted. Several were allocated to the main development team as test vehicles. Two were used for brake testing in Scandinavia, at least three were barrier-tested during July 1980, and the Styling Department used at least one and supposedly two of them for styling trials. Some were scrapped. But just as the French project was winding down, so the possibility of another major military contract came up. This time, it was for the Swiss Army.

The Swiss wanted five test vehicles, two short-wheelbase and three long-wheelbase types. All were to be powered by 4-cylinder petrol engines, as the Swiss did not want the V8, and all were to have automatic gearboxes. At least one of each type was to be a Station Wagon. So as a first step,

This 100in military prototype had RHD and a more orthodox pick-up specification. The grille and badge were Series III Stage 1 V8 types and show the oil cooler clearly.

Bob Seager's team completed two of the vehicles originally intended for the French trials to meet the Swiss requirement. Both had the latest experimental 2.5-litre enlargement of the 4-cylinder petrol engine; one was a four-door soft-top with a 5-speed manual gearbox, while the other was a four-door Station Wagon with a Borg Warner 3-speed automatic

This was another Swiss oddity, although the body configuration was familiar. Again on the 100in wheelbase, it is a four-door Station Wagon, built mainly from modified production panels. The power train, however, consists of a 2.5-litre diesel engine driving through an automatic gearbox.

gearbox. The Station Wagon body was built for Land Rover by Jensen at West Bromwich.

The Swiss trials began sometime around June 1979 and lasted over a period of nearly five years: the One Ten trial vehicles were not delivered until 1984. Bob Seager remembered that the 2.5-litre engine suffered from fuel vaporization problems and that the 3-speed automatic was forever in the wrong gear and overheated as well. Even so, the Swiss were satisfied with the performance of the two 100in prototypes, were therefore looking forward to receiving the three 110in models for test in due course.

One of the other contenders for the Swiss contract was again the Mercedes-Benz G-Wagen; at this point, as Bob Seager understood it, the German company exercised a little muscle. He believed that they managed to get negotiations delayed until Swiss law changed to demand catalytic converters on all new vehicles. Mercedes already had a 4-cylinder petrol engine with a catalytic converter, but Land Rover did not.

So, in order to stay in the race, Land Rover had to switch from the 2.5-litre 4-cylinder engine to its only other option, which was the 3.5-litre V8 that the Swiss had initially not wanted. Although a version of this with catalytic converters was in preparation for a number of markets, there was no 24V variant of it and the Swiss wanted 24V vehicles. So a distributor was specially modified to suit and the Swiss vehicles – presumably the three originally called for, although no trace has been found of a Station Wagon – were delivered for trials. These all had the 4-speed ZF gearbox that had replaced the unsuccessful Borg Warner type in the specification.

That modified distributor proved a weak link. Poor waterproofing let it down during the trials, with the resulting misfire destroying the catalytic converter, which of course raised exhaust emissions to levels unacceptable to the Swiss. Meanwhile, it looks as if work had continued on perfecting the Swiss 100in, and at least one more vehicle was built with a V8 engine, ZF automatic gearbox and limited-slip differentials in both front and rear axles. It also had a special 24V to 12V converter in the electrical system, which came into play during the engine shut-down cycle. This was almost certainly designed as a solution to the problem raised by the lack of a 24V distributor.

In the end, Land Rover lost the contract. In some ways, it was a mercy because the 100in model had disappeared from Land Rover's product plans long before the collapse of the Swiss endeavour. In its place had come a proper short-wheelbase model, which entered production in 1984 as the Ninety. If the Swiss contract had gone ahead, Land Rover would have been obliged to build a batch of 100in vehicles with major differences from the standard production models. The prestige would have been great, but so would the cost.

110 FEPs and Pilot Production, 1979–83

Just before the Swiss diversion, the Stage 2 project team had started work on a large batch of Fully Engineered Prototype (FEP) 110in models. These vehicles had a variety of specifications to try out all the planned variants of the 110in Stage 2 model and there seem to have been forty-eight of them,

The Styling Department tried different colour schemes on a number of early One Tens. This one was eye-catching, to say the least, but was not adopted for production. The black wheel centres looked good, but would have added to production costs.

The production-style slatted grille simply was not available when this picture was taken to show the truck cab variant of the One Ten. The tropical roof for the cab did become available for hot-climate territories, but the military bumpers here were probably just what came to hand when the vehicle was being built up.

This is another early iteration of a One Ten, in this case a Station Wagon. Interesting features being tried out here are the headlamp washers (expected to be a requirement in some countries) and the tropical roof (which was not needed, thanks to a more efficient type of headlining). Again, the grille is a Stage I V8 type, because the production style was not yet available.

built between May 1979 and about February 1980. Under development at the time for introduction in autumn 1980 were five-bearing versions of the 2.25-litre 4-cylinder petrol and diesel engines, and those FEP vehicles built as diesels certainly had prototypes of the new engine. It is probable that the FEPs with 4-cylinder petrol engines also had prototypes of the new version.

The FEP 110in models were numbered from 110F1 upwards, the F presumably being short for FEP, although the letters were used in full on other vehicles, including the final

This is the 4-cylinder diesel engine in a One Ten prototype, pictured while the vehicle was being hand-assembled. Note the label reading 'Sample' on the steering wheel centre cover at bottom right.

Some of the FEP One Tens were used for trial installations of optional equipment, in this case a front-mounted electric winch. The starting-handle bracket on these early models is very apparent here.

Swiss 100in prototype. Those which received British registration plates mostly had numbers within the HAC … V and HWK … V series; three were registered later with OKV … W numbers. Most were scrapped in the early 1980s after being run into the ground during tests, but a small number were sold off through Land Rover dealers.

The launch specification was signed off some time in 1982 and the next stage was to build the production prototypes. There were at least fifty-six of these, again with specifications embracing the full variety planned for production, and they had standard production chassis numbers. Most of them were registered in the UK between November 1982 and April 1983,

ABOVE: **Another very early One Ten was this one, built as a LHD window hardtop with the Series III windscreen and Stage I grille. It still survives, in the Dunsfold Collection.**

LEFT: **The pilot-production models were mostly registered with CWK ... Y numbers, and they became the launch fleet. CWK 53Y was a hardtop and this picture was issued to the press when the One Ten was new.**

in the sequence CWK 2Y to CWK 59Y (some numbers in the sequence went to other Land Rover vehicles). Many of these were used for the press and dealer launches of the new model in March 1983, but some were sent to Special Projects for conversion, while others became working vehicles for departments such as the Land Rover Demonstrations Team. Another small batch, registered as CWK 82Y to CWK 85Y, went straight to the Engineering Department in December 1982.

RIGHT: **Some of the pilot-build vehicles were used as the basis of early conversions. CWK 14Y became the prototype Backhoe Digger, although the conversion did not become available until the mid-1980s.**

LEFT: **Demonstrating its axle articulation during a launch event at Solihull, this is CWK 8Y, which was a standard Station Wagon.**

BELOW: **CWK 42Y was also fed into the conversions programme, becoming the prototype Shorland Series 5 armoured patrol car.**

LEFT: **Some of the CWK … Y models were pressed into service by Land Rover itself. CWK 6Y was one of several that belonged to the Demonstrations Team. The team's all-white colour scheme could not be matched from the standard paint options for many years.**

BELOW: **The CWK … Y fleet lines up at Solihull for the 1983 launch event. This was the 'road' fleet, which remained fairly clean; other CWK … Y vehicles formed a dedicated off-road fleet, which did not.**

Ninety and One Ten Crew Cab, 1981–3

That batch of CWK … Y models was critical to the engineering programme in more than one way. It not only provided the engineers with vehicles to near-production standard that they could test to their hearts' content; it also provided vehicles that could be modified and developed in connection with the next stages in the engineering programme. And the most important next stage in the engineering programme was the development of a short-wheelbase model.

Mike Broadhead had taken over from Tom Barton as Chief Engineer in 1981, and he favoured a 90in (2,286mm) wheelbase rather than the 100in (2,540mm) type that was already undergoing serious development for military purposes. So a special chassis with a 90in (2,286mm) wheelbase was con-

RIGHT: **The first Ninety prototype was registered in the CWK ... Y pilot-production batch and it really did have a 90in wheelbase. Chief Engineer Mike Broadhead thought it was too small.**

BELOW: **The second Ninety prototype was again converted from a CWK ... Y One Ten (*left*). The rear body panels were from a Series III 88in model and were too short, so a special filler panel had to be made up (*right*).**

structed at the Bordesley Green plant and was completed with a Station Wagon body in December 1982. It took a registration number from the CWK ... Y sequence then being allocated to the pilot-build One Tens, and became CWK 30Y.

However, Mike Broadhead thought it did not show enough improvements over the Series III 88in (2,235mm) models then in production, so he had a second prototype built on a slightly longer wheelbase to incorporate a longer load bed and make room for an adjustable driver's seat. This became CWK 40Y and its wheelbase worked out at 2,360mm (Land Rover had been working to metric dimensions for several years), which translated as 92.9in. To save time, this prototype was constructed by shortening a pilot-production One Ten chassis and in fact all nine subsequent short-wheelbase prototypes were made in the same way. The last prototype of what Land Rover's marketing people wanted to call the Ninety – despite its 92.9in wheelbase – was built in September 1983 and the new model was ready for production by the summer of 1984.

Then, before the new 110in coil-sprung model was launched, Land Rover also came up with a third model that had not been in the original plan. This was an ultra-long wheelbase derivative, and its prototype was built from one of the CWK … Y pilot-build vehicles.

In fact, the prototype of the ultra-long-wheelbase model had been completed by the time of the One Ten's launch in March 1983 and many members of the media saw it at that stage. It had originated a couple of years earlier, when Product Planning Manager Mike Gould had been in Australia attending a conference for those manufacturers who planned to tender for the forthcoming Australian Army vehicle contract, known as Project Perentie. Struck by the popularity of Japanese double-cab pick-ups in Australia, Mike thought that Land Rover would be able to build a similar vehicle out of the forthcoming One Ten.

So he created one on paper by cutting out profile drawings from a One Ten technical manual and pasting them together. The front section of the cab was standard One Ten; the rear section was the back of a truck cab, with the rear side door from a Station Wagon modified to suit the straight rear edge of the truck cab's door aperture. Behind that, he added a High-Capacity Pick-Up (HCPU) back body. The wheelbase measurement, dictated by the components used, ended up at 135in (3,429mm).

Marketing Manager Steve Westwood liked the idea and gave it his support. However, when the idea reached Solihull's engineers, they immediately pointed out that a wheelbase of this length would need a two-piece rear propshaft, which would be expensive. The solution was to base the dimensions on the longest propshaft then available off the shelf from supplier Hardy Spicer, and with that in place the wheelbase ended up as 127in (3,226mm). Cutting the HCPU back body down to fit was not difficult, although it meant that the spare wheel had to be relocated onto the bonnet. So the job of making a feasibility prototype was turned over to Spencer Abbott in Birmingham, along with a pilot-build One Ten, CWK 55Y.

The vehicle was built up with a five-seater 'crew cab', as Mike Gould had envisaged. And because of the engineering programme quirks already outlined, the so-called One Ten Crew Cab was introduced before the Ninety.

CWK 55Y became the prototype One Ten Crew Cab on an extended-wheelbase chassis. The name One Two Seven would later be used for production models.

CHAPTER FOUR

THE ORIGINAL RANGE, 1983–1986

This cutaway drawing was issued when the One Ten was launched in 1983, to show the vehicle's basic layout. This is a Station Wagon with the V8 petrol engine.

One Ten, One Two Seven and Ninety, 1983–6

Preparing for the launch of the new Land Rover demanded a great deal of thought at Solihull. Although there had been incremental upgrades to the original Land Rover ever since 1948, there had never been a redesign that was as far-reaching as the introduction of coil springs. This was one probable reason why the new models were not marketed as Series IV types, the name that they might logically have taken as it was next in the existing sequence.

Instead, the marketing team decided that the models should in effect be rebranded and the new 110in (2,794mm) type should have a new style of name. So it became a Land Rover One Ten in publicity material and in Land Rover literature, even though its grille badge always read 'Land Rover 110', with the name in figures. The name was certainly more memorable than 'Hundred and Ten' or 'One One O', which would have been the obvious alternatives. It also allowed the new short-wheelbase model to be branded as a Ninety, even though its grille badge would again carry the name in numbers.

The model introduction strategy was a new one for Solihull, too. Whereas the short-wheelbase and long-wheelbase

ABOVE: **The inner wings had been designed to give working space around the engine; see the prototype on page 24 for comparison. Even so, the 2.25-litre petrol engine looked a little lost in such a large engine bay and needed a huge cowling between fan and radiator.**

RIGHT: **These were some of the critical elements of the new One Ten: coil springs and, on the front wheels, disc brakes.**

derivatives of earlier major model changes had been introduced together, the coil-sprung models would have a more protracted arrival. Only the One Ten would be introduced at first, and it would replace the Series III 109 models in a rolling programme that would last just over two years. This would allow the new coil-spring technology to gain acceptance with Land Rover's more conservative customers; Land Rover itself claimed that some overseas fleet buyers would test a vehicle for as long as three years before placing a bulk order and in the meantime it was important to continue production of a vehicle that such customers had already tried and tested.

The Ninety would follow a year or so later, and in the meantime the Series III 88in (2,235mm) would continue to be built for those who wanted short-wheelbase models. In practice, the One Ten prepared the ground so effectively for the Ninety that there was no need to continue building Series III 88s after the summer of 1984 and the model was withdrawn. The final Series III 109s were nevertheless not built until summer 1985, as there were large fleet orders from overseas and from the British military to be fulfilled.

As for the One Two Seven models, these were treated in the early days as conversions of the One Ten rather than as separate models in their own right, and therefore they were not seen as critical to the overall marketing strategy. In practice, sales began at the end of 1983 but were very slow until mid-1985, when the One Two Seven became a model in its own right.

Up to the middle of the 1980s, the new coil-sprung models continued in the tradition established by the leaf-sprung types. But from mid-1985, it was clear that things were beginning to take a new direction. There was a new emphasis on the Station Wagon models to exploit the fashion for using 4×4s as everyday vehicles and the fruits of the company's new focus on European markets became very apparent.

The 1983 Model Year

Land Rover's new Land Rover, as publicity material called it, was introduced to the public at the Geneva Motor Show in March 1983, although dealers had been briefed a month or so earlier. In the beginning, the One Ten was available only in the UK and Switzerland, although it was introduced to other markets gradually over the next few months.

The model line-up was at least familiar and there were five body styles available. These were the soft-top, hardtop, truck cab pick-up, High-Capacity Pick-Up and Station Wagon; there were also better-equipped County Station Wagon derivatives. The One Ten Station Wagons normally came with twelve seats for the UK, where tax regulations had favoured this seating capacity since the early 1960s. For other markets, however, the Station Wagon norm was ten seats, with four individual seats replacing the two bench seats in the rear that theoretically provided room for six people.

These first models were available with a choice of three engines, two petrol-fuelled and one diesel, and all of them uprated from their Series III incarnations. Least powerful was the 4-cylinder 2.3-litre diesel, with overhead valves and a swept volume of 2,286cc. Much earlier, this had been known as a 2¼-litre engine; the dimensions had not changed, but it had been rebranded when re-engineered in 1980 to take a five-bearing crankshaft in place of the original three-bearing type. It had indirect injection, maximum power of 60bhp and a torque peak of 103lb ft at 1,800rpm. In One Ten form, it had a key-operated (solenoid) cut-out for the first time.

The smaller of the two petrol engines shared the 4-cylinder overhead valve (OHV) layout of the diesel and was again a 2.3-litre type with five main bearings and a 2,286cc swept volume. For the One Ten, it had been improved with redesigned inlet and exhaust manifolds, a Weber 32/34 DMTL carburettor in place of the earlier Solex and a redesigned camshaft. The main aim of the changes had been to improve low-speed torque, but there had been power gains as well and the latest 2.3-litre engine had 74bhp at 4,000rpm and 163lb ft of torque at 2,000rpm. (Note that these figures were measured by SAE standards and that Land Rover changed a few years later to the DIN standard, which was generally stricter.)

The most powerful engine was the 3.5-litre petrol V8, now with 114bhp at 4,000rpm (25 per cent more than was available on Series III Stage I V8 models) and 185lb ft of torque at 2,500rpm. This engine was expected to prove most popular in Station Wagon models because it offered a combination of refinement, flexibility and low-speed lugging power, plus the ability to cruise at 80mph (129km/h) or more on motorways – as long as high fuel consumption was not an issue.

The V8 models came with the 4-speed LT95 gearbox already used in the Range Rover and this had its own integral two-speed transfer gearbox. However, the 4-cylinder petrol and diesel engines drove through the LT77 5-speed gearbox, originally developed for Rover and Triumph cars in the mid-

This was the dashboard of a 4-cylinder One Ten County Station Wagon in 1983. The steering wheel now resembled the Range Rover type and the whole dashboard and instrument panel seemed much more sophisticated than their Series III equivalents. The seats have cloth upholstery, of course, which was standard with the County specification.

LEFT: **The horizontal stripes – actually a decal – show that this is a County Station Wagon. The vehicle is in fact CWK 37Y, one of the pilot-production models used on launch duties.**

ABOVE: **Permanent four-wheel drive would become standard, but not just yet; it was possible to order 4-cylinder One Tens with the old Series III 76mm gearbox and transfer box with selectable four-wheel drive.**

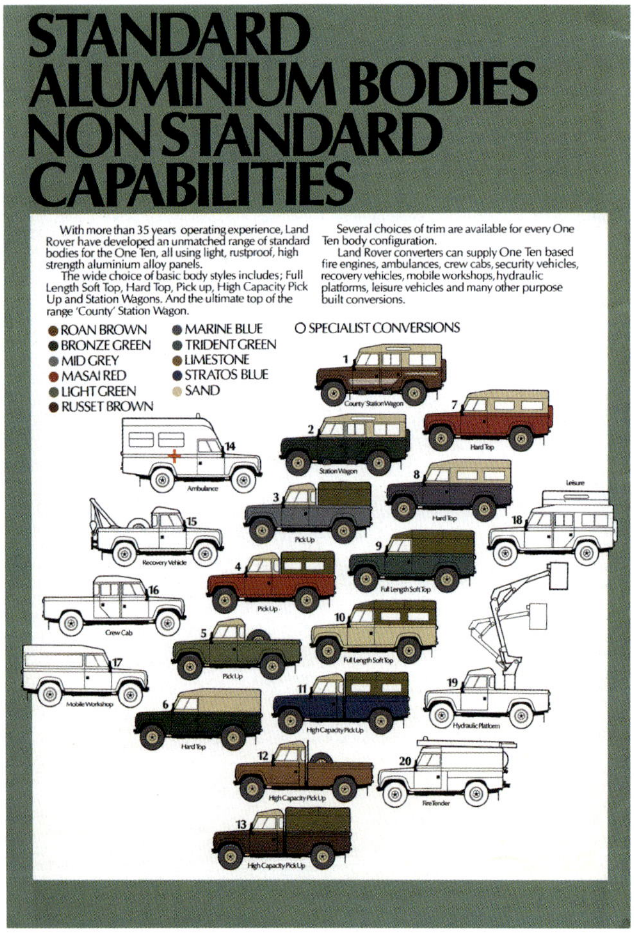

LEFT: **There was a wide range of options from the very start. This page from the 1983 sales brochure shows that the One Ten Crew Cab (no 16) was already on offer – but that there were plans to change to an angled edge for the rear door. Note also how these early models all had Limestone upper panels.**

1970s and now accompanied by the separate LT230R two-speed transfer gearbox that had first been seen in Range Rovers in autumn 1982.

The One Ten was an immediate success, with Land Rover claiming at the end of 1983 that it had accounted for almost a third of home-market Land Rover sales that year. The biggest success belonged to the County Station Wagons, whose sales exceeded Land Rover's own expectations. They rode the crest of a wave of interest in 4×4 vehicles for everyday transport that would characterize the rest of the 1980s.

County Station Wagons were readily recognizable by a set of decal side stripes accompanied by a script 'County' badge on each side at the rear and on the tail panel. Like the standard Station Wagons, they did not have the 'tropical' roof that had characterized earlier Station Wagons of all kinds. Instead, a new headlining made of resin-impregnated felt provided all the heat-insulating properties necessary.

Special features also justified the higher prices asked for the County models and although many of their special features could also be found on the options list for other models, there were some that were not. County features included tinted glass, halogen headlamps, a reversing lamp, mud flaps both front and rear, side repeater flashers on the front wings, a hazard warning system, a PVC spare wheel cover and a bonnet lock. Power-assisted steering, self-levelling rear suspension and radial tyres were also standard; other models still came with cross-ply tyres. On the inside, County models had brown cloth upholstery instead of the otherwise universal black PVC, rubber pedal pads, a dipping interior mirror, a clock, a voltmeter, an ash tray and sun visors.

A wide range of optional equipment was available, as had always been the case with Land Rovers. For the One Ten, there was the usual array of commercially oriented extras such as winches and power take-offs, but there were many other types of extras. These were available both individually and grouped into Option Packs covering Electrical, Protection, Towing and Interior Appointment additions.

There was one mildly amusing mishap as the One Ten came on stream. The new sales brochures for Land Rover's new Land Rover were dated February 1982 – a year out of date.

The 1984 Model Year

The concept of changes for every model year was still a relatively new one for Land Rover. Although the company had embraced the need for it wholeheartedly on its Range Rover models, it took some time to do the same for the utility Land Rovers. So the start of the 1984 model year was not clearly defined as far as the One Ten was concerned, even though October 1983 marked the start of the 1984 model year for Range Rovers. For One Tens, the 1983 model year continued until the end of the 1983 calendar year, or thereabouts, and the 1984 model year started somewhere around the time that 1983 slipped into 1984. Its start was defined, if rather imprecisely, by the introduction of the new ultra-long-wheelbase model at the end of 1983 and by the replacement of the 2.25-litre diesel engine by a 2.5-litre derivative in February 1984.

Land Rover had already received its first order for the ultra-long-wheelbase model by the time of the One Ten's launch

The One Ten Crew Cab began life as a hand-converted model. This very early one was ordered by the National Rivers Authority and shows very clearly the original configuration with a straight trailing edge for the rear door.

in March 1983; long-standing customer Southern Electricity wanted a small batch. However, there were delays in getting the model homologated, so even though it was pictured in the first sales brochures in early 1983, it did not become available until the end of the year. It was not built on the main assembly lines, but was converted from a standard One Ten by the Special Installations Department. Chassis were cut in half and placed in a special jig, then a 17in (432mm) extension was welded in place and the new body built up on top of it. This was a slow and labour-intensive process, which meant that production volumes were very low in the early days.

As released to the public in late 1983, the new model carried the name of One Ten Crew Cab. This reflected its origins as a double-cab pick-up, but the production models dispensed with the Land Rover 110 grille badge, which was clearly inappropriate for their additional size. Instead, they had the printed silver 'Land-Rover' badge that had been used on the Series III Stage 1 V8 models between 1979 and 1983. It was an anachronism of course as Land Rover had stopped using the hyphen in its name at the end of the 1970s.

Before long, it became apparent that there were other potential uses for the 127in (3,226mm) chassis. A self-levelling unit was added to the rear axle to allow for higher payloads and, as production volumes built up, so assembly became more slick and began to use properly tooled parts rather than individually fabricated items. However, it would be mid-1985

Nobody talked about this one at the time, but a special One Ten Station Wagon was delivered to the Royal Family for use at Balmoral. The all-over green colour scheme with black bumpers was unique, as were the grab handles on the door pillars; these were designed so that the ghillies could ride on the side steps of the vehicle when the royals were out shooting on the estate.

STANDARD COLOURS AND TRIMS — 1983 and 1984 Model Years

Paints

Bronze Green	LRC001	
Light Green	LRC005	
Limestone	LRC007	
Marine Blue	LRC006	
Masai Red	LRC378	County models only
Mid Grey	LRC008	
Roan Brown	LRC324	County models only
Russet Brown	LRC318	County models only
Sand	LRC004	
Stratos Blue	LRC327	County models only
Trident Green	LRC325	County models only

Roofs, Wheels and Soft-Tops

Roofs, wheels and upper side panels were universally painted Limestone (LRC007). Soft-tops were all Khaki.

Side Stripes

County models had Limestone coloured stripes with the County name in script on the rear side panel.

Upholstery and Trim

Standard upholstery was Black vinyl. County models had County Brown cloth upholstery.

CAMEL TROPHY VEHICLES

Between 1981 and 1998, Land Rover was the primary sponsor of the Camel Trophy adventure competition. In each of those seventeen years, the company provided a fleet of vehicles for the international teams to use – the vehicles were One Ten Station Wagons in 1984 and Nineties in 1985 and 1986.

A posed image, perhaps, but this picture from the 1984 Camel Trophy helps to illustrate the excitement that always accompanied the event.

The Camel Trophy was always good for some dramatic photography and this picture from the 1986 event is fairly typical. It all helped to build the Land Rover legend and create an image of toughness that helped to sell vehicles.

The 1984 event was held in the Amazon Basin in Brazil, with teams from twelve countries. The One Tens all had the 2.3-litre diesel engine, as the 2.5-litre type was too new to allow preparation and training before the event. For the 1985 event in Borneo, the team vehicles were 2.5-litre diesel Ninety Station Wagons. Heavy rains dogged this event and at one point the whole convoy had to be rescued by helicopter. The 1986 event took place in Australia and the fourteen teams drove diesel Ninety Station Wagons again.

before homologation was completed and the 127in chassis could move on to the next stage of its career.

The new diesel engine appeared just a few months later and was announced at the Amsterdam Show, which opened on 4 February 1984. It actually represented the next stage in the engine development programme after the switch to five main bearings from three for the 1981 model Series III vehicles, and its increase in capacity had been achieved by lengthening the stroke. Power went up to 67bhp at 4,000rpm, an increase of 12 per cent over the older engine's 60bhp, and torque went up to 113lb ft, a 10 per cent improvement over the older engine's 103lb ft.

There was more to it than that, though, because the engine had been redesigned with a modern toothed rubber-belt drive for the camshaft in place of the original roller-chain drive. This belt also drove the 2.5-litre engine's new DPS injector pump, which promised more accurate fuel metering and longer retention of tune than earlier types. In that way, it was expected to reduce fuel consumption. The DPS pump also benefited from a self-priming system instead of the hand primer of earlier types, which reduced the need for owners to get their hands dirty under the bonnet. The front-end redesign associated with the new timing belt gave Solihull's engineers the opportunity to make other worthwhile changes and they were able to save a little weight by using alloy castings for the new front cover and water pump housing. Cold-starting was improved, too, as the 2.5-litre engine had sheathed-element glow plugs.

'There will not be a 2½-litre petrol engine,' stated the Product Bulletin that Land Rover sent out to its dealers in December 1983 to advise them of the new engine's introduction. The thought was badly expressed: what the company meant was that no 2.5-litre petrol engine would be made available just yet. There was certainly one in the development programme.

The 1984 model year ran only to June 1984, because in that month Land Rover made its 1985 models available. There was a facelift for the One Ten and, most importantly, the Ninety became available as a replacement for the old Series III 88in model.

The 1985 Model Year

The 1985 model year was characterized above all by the introduction of the Ninety, which was initially available only with the 2.5-litre 4-cylinder petrol and 2.25-litre diesel engines. The new short-wheelbase model's arrival was timed to coincide with a facelift for the One Ten and so benefited from all the new features. Both models were introduced in May 1984 at a media event held at Eastnor Castle. Showroom sales began in June, but few Ninety models found owners before the autumn.

The most important element in that 'facelift' was a change on all models from the sliding windows in the front doors to conventional winding windows. These doors, originally designed by Santana in Spain, made a big difference. Despite appearances, they were still made in two sections, but the join between the two was concealed under a galvanized capping. Not only did they allow the use of glass instead of scratch-prone Perspex, they also made possible continuous door seals, which reduced wind noise and improved the sealing of the interior against dust and water. Redesigned interior door trims now incorporated sill-locking buttons for the first time on a Land Rover.

Equipment upgrades were made across the range. All the 1985 models had locks for the bonnet, steering column and fuel filler, with a cigarette lighter and dipping mirror becoming standard as well. To these improvements, all Station Wagons added a heated rear window and a rear wash-wipe system, while the County Station Wagons now came with carpets in the passenger cabin and a radio preparation kit of two speakers and an aerial – although the actual head unit still cost extra, at least in the UK. County models had side repeater indicators as standard, too.

The big news for the 1985 model year was the arrival of the new short-wheelbase model. This is a very early Ninety in truck-cab guise. The Limestone paint was now confined to the roof and the upper panels were in the main body colour.

There was also a cosmetic change for the County Station Wagons, which now had a new style of side decal bearing the County name on the doors. The new decals were designed to suit both One Ten and Ninety models, with a trailing design at the rear that simply dispensed with one section on the shorter models. They were also matched to the main body colour. Soft-tops on workhorse models were now Matt Stone instead of the earlier Khaki, which was nevertheless retained for military models.

The Ninety came as standard with 6.00 × 16 tyres and the usual factory fit was Avon Traction Mileage types. Strong demand in the UK persuaded Solihull to give UK dealers priority in the build schedules for Nineties between January and March 1985. There were still leaf-sprung Series III 88s to be had, but few buyers were tempted; the new Ninety was a better proposition all round, even if it was more expensive.

However, some buyers probably held off, sensing that a V8-engined Ninety could not be long in coming. They were right: production began sometime around March 1985 and the new model was announced on 7 May. It was seen as a late 1985 model rather than an early 1986 model and with it came several improvements that affected other utility Land Rovers as well. Some were minor, such as new pedal pads, better-spaced pedals on left-hand-drive vehicles and a new integral ash tray on the dash instead of the earlier bolt-on type. But the major changes affected the power train.

All V8 engines, whether for One Tens or Nineties, now had electronic ignition like that already in use on the Range Rover. For most countries outside Europe, the V8's power was also increased to 134bhp. However, Europe and the UK remained with the existing 114bhp tune, while Australia remained with its unique 125bhp tune.

With the Ninety V8 – and the latest One Ten V8s as well – came a long-overdue 5-speed gearbox to replace the noisy old 4-speed LT95 type. This was called the LT85 (because there were 85mm (3.3in) between the centres of the main-shaft and the layshaft) and even though it carried the LT designation associated with Leyland Transmissions, it had actually been designed by Land Rover's Santana licensees in Spain. It had originally been designed as a heavy-duty gearbox for use with Santana's own 6-cylinder engines. The overdrive top gear helped to reduce the V8's heavy fuel consumption a little and the LT85 gearboxes came with an uprated transfer gearbox

There was, of course, a Ninety Station Wagon as well, seen here in County trim. The new County stripes were shared with the One Ten, which simply had a longer version of them.

LEFT: **White now became available as a production option, creating a particularly striking vehicle when combined with the blue County decals.**

BELOW: **Two of the Ninety launch fleet were pictured at Land Rover's off-road centre at Eastnor, with the castle in the background. Side indicator repeaters were standard on UK-model Counties for 1985, but not on utility models.**

The change from Limestone to body-coloured upper panels made a surprising difference, as this One Ten County Station Wagon shows.

called the LT230T, which also had reduced gate loads for the High–Low change.

However, the 4-cylinder models of both wheelbases stayed with the LT77 5-speed gearbox and LT230R transfer box. The 4×2 option, with the old 4-speed gearbox, now disappeared from the options list and the range went over completely to a permanent four-wheel-drive configuration. The Ninety models also had taller high-range gearing in their transfer boxes than the One Tens; their lighter weight allowed them to pull this with ease.

Four-cylinder Nineties, supplied as standard with Avon Traction Mileage tyres up to this point, now started to appear with Goodyear Hi-Miler S&G tyres as Solihull switched to dual suppliers. However, the Ninety V8 had its own tyre specification and came with 205R16 radials as standard. A 7.50 × 16 size was an option. There were more changes introduced on production in July: One Ten County models switched to Avon Rangemaster tyres and all One Ten V8s switched to Pirelli LR04 7.50 × 16s. The other models continued with Avon Ranger 7.50 × 16s.

From April 1985, Land Rover also began a new drive to expand its Approved Conversions range through the newly reorganized Special Vehicles division (*see* Chapter 1). Meanwhile, it was in November 1984 that the One Ten was introduced for the Australian market, with the variants here including some that were particularly interesting. To meet domestic demands that were mostly driven by the availability of large diesel engines in Japanese 4×4s, Land Rover had been fitting

By this stage, the switch had been made to an angled rear door edge for the Crew Cab model, which also took on the name of One Two Seven.

Isuzu 3.9-litre 4-cylinder diesel engines instead of Land Rover's own 2.25-litre diesel into Series III models since 1981. So the One Tens that were sold in Australia were hardtops and Station Wagons, with this engine as the diesel option and the 3.5-litre V8 as the petrol option. The V8 models were imported from the UK as fully built vehicles, but the diesels were assembled in Australia from CKD kits at the Moorebank assembly plant in Sydney's suburbs.

Australia also introduced more special derivatives at the start of the 1985 model year. There was an extended-wheelbase One Ten with 119.9in (3,046mm) between axle

STANDARD COLOURS AND TRIMS

1985 Model Year

Paints

Bronze Green	LRC001
Roan Brown	LRC324
Ivory White	LRC354
Slate Grey	LRC348
Marine Blue	LRC006
Stratos Blue	LRC327
Masai Red	LRC378
Trident Green	LRC325

All colours were available on County models.

Roofs, Wheels and Soft-Tops

Roofs and wheels were always painted Ivory White (LRC354). Soft-tops were Matt Stone.

Side Stripes

County models had Limestone and Beige or two-tone Blue coloured stripes with the County name picked out on each front door.

Upholstery and Trim

Standard upholstery was Black vinyl. County models had County Brown cloth upholstery.

centres that was offered as a chassis-cab with either engine; though always officially a One Ten chassis-cab, this was soon rechristened the 120 by customers in honour of its near-120in wheelbase. The diesel-engined version of this was also offered ready equipped with a uniquely Australian 'tray' back body made out of aluminium by Alcan.

On top of this came the uniquely Australian 6×6 model. As Chapter 11 shows, this was a civilian spin-off from the special two-tonne Land Rover developed for the Perentie military contract; Chapter 11 also describes the model in more detail. (At this stage, the military contract had not yet been awarded, but Land Rover did win it in 1986.) The 6×6 entered production with a turbocharged derivative of the Isuzu diesel as the standard engine, and with the naturally aspirated Isuzu engine and the 3.5-litre V8 petrol engine as special-order alternatives.

The 1986 Model Year

The 2.5-litre petrol engine that Land Rover had promised would not be introduced put in its appearance in August 1985, for an October start to the 1986 model year. Though sharing its longer stroke with the 2.5-litre diesel engine, it did not share that engine's toothed-belt camshaft drive, but retained the original chain-drive configuration. The 2.5-litre

The One Ten County was by now a handsome beast and needed no change for the 1986 model year.

petrol engine completely replaced the 2.3-litre type, bringing with it greater power and torque.

The figures, expressed to the DIN 70020 Net standard that Land Rover now embraced as a policy, were 83bhp at 4,000rpm and 133lb ft at 2,000rpm. Comparing them with the figures normally quoted for the earlier 2.3-litre engines is impossible, because these were always measured by different standards. For what it is worth, though, those figures had been 74bhp at 4,000rpm and a suspiciously high 163lb ft at 2,000rpm. The new figures for the 2.5-litre diesel engine were 68bhp at 4,000rpm and 117lb ft at 1,800rpm, while those for the 3.5-litre petrol V8 were 114bhp at 4000rpm and 185lb ft at 2,500rpm.

There were some cosmetic changes for 1986, too. Not immediately obvious was a cleaner side profile that came from the deletion of galvanized capping rails below the door windows. These were no longer needed, as the doors were now proper one-piece types. In Europe, the side repeater indicators introduced on County models for 1985 now became standard wear across the range.

The other major change for the 1986 model year affected the One Two Seven models, which were now fully homologated. Their assembly was now carried out on the main production lines rather than by the Special Installations Division, which enabled production volumes to be increased. There was also a redesign of the Crew Cab rear doors, which was necessary to allow assembly on the main lines. The original vehicles had a straight trailing edge to this door, which required a special frame and panel not used on other models. From the start of the 1986 model year, an angled panel was used at the bottom rear of the opening so that standard Station Wagon rear doors could be fitted.

There were also some minor running changes in February 1986. A fuel filler cap with integral barrel lock became standard and was also used for an auxiliary tank when one was fitted. Towing rings were deleted from the optional Towing Pack and the new recommendation was to use the chassis-mounted towing and lashing eyes. The recommended oil for the LT77 5-speed gearbox was also changed to Shell Donax TF from the earlier Dexron II ATF.

ABOVE: **Carefully posed, but in another split second the moment would have been lost ... this One Ten Station Wagon became a 'Follow Me' vehicle at London's Heathrow Airport and is pictured with a British Airways Concorde supersonic airliner in the background.**

LEFT: **Air conditioning was an extra-cost option and with it came a condenser radiator that sat ahead of the main radiator and pushed the front panel forwards. The bulkhead vents were permanently closed and were finished in black, too. The vehicle here is older than it has been made to appear: it still has the two-piece doors discontinued with the 1985 model year.**

The 1986 Land Rover Ninety Hurricane (The Netherlands)

Land Rover was not in the habit of creating special editions in the mid-1980s; that came later, as Chapter 10 shows. However, there seems to have been an exception in the Netherlands, and that was the Ninety V8 Hurricane model.

Little information is available about this special edition, which was supposedly of twenty-five vehicles. The story attached to it is that the Dutch Navy ordered a batch of Ninety V8 Station Wagons, but that the order was cancelled at the last minute. Left with these Station Wagons, painted a non-standard dark blue, Land Rover Netherlands made the best of a bad job and turned them into a 'limited edition' by adding side stripes with the same pattern as those on a Ninety County V8, but with the word 'Hurricane' replacing the word 'County'. They had inward-facing bench seats in the rear rather than the standard four individual seats and are said to have had Rostyle wheels when new.

Land Rover Netherlands have not confirmed this story, but the reason may be because of a company-wide blanket refusal to comment on military orders.

NOTE: For Vehicle Identification Numbers and Engine Identification Numbers applicable to the 1983–6 period, *see* the consolidated lists for 1983–90 in Chapter 5.

The Hurricane was a special edition created out of necessity in the Netherlands. This example has been fitted with aftermarket wheels, tyres and side-runners.

STANDARD COLOURS AND TRIMS — 1986 Model Year

Paints

Arizona Tan	LRC341
Slate Grey	LRC348
Bronze Green	LRC001
Stratos Blue	LRC327
Ivory White	LRC354
Trident Green	LRC325
Marine Blue	LRC006
Venetian Red	LRC440

All colours were available on County models.

Roofs, Wheels and Soft-Tops

Roofs and wheels were always painted Ivory White. Soft-tops were Matt Stone.

Side Stripes

County models had Limestone and Beige or two-tone Blue coloured stripes with the County name picked out on each front door.

Upholstery and Trim

Standard upholstery was Black vinyl. County models had County Brown cloth upholstery.

SPECIFICATIONS

NINETY, ONE TEN and ONE TWO SEVEN models 1983–86

Petrol Engines

Type	2.3-litre
Bore and stroke	90.47mm × 88.9mm
Capacity	2,286cc
Valves	4-cylinder with ohv
Compression ratio	8:1
Carburettor	Single Weber 32/34 DMTL twin-choke carburettor
Max. power (DIN)	74bhp at 4,000rpm and 120lb ft at 2,000rpm

Type	2.5-litre
Bore and stroke	90.47mm × 97mm
Capacity	2,495cc
Valves	4-cylinder with ohv
Compression ratio	8:1
Carburettor	Single Weber twin-choke carburettor
Max. power (DIN)	83bhp at 4,000rpm and 133lb ft at 2,000rpm

Type	V8
Bore and stroke	88.9mm × 71.1mm
Capacity	3,528cc
Valves	8-cylinder with pushrod ohv
Compression ratio	8.13:1
Carburettor	Two Zenith-Stromberg carburettors
Max. power (DIN)	114bhp at 4,000rpm and 185lb ft at 2,500rpm

Diesel Engines

Type	2.3-litre
Bore and stroke	90.47mm × 88.9mm
Capacity	2,286cc
Valves	4-cylinder with ohv and indirect-injection
Compression ratio	23:1
Fuel system	Indirect injection
Max. power (DIN)	67bhp at 4,000rpm and 103lb ft at 1,800rpm

Type	2.5-litre
Bore and stroke	90.47mm × 97mm
Capacity	2,495cc
Valves	4-cylinder with overhead valves and indirect-injection
Compression ratio	21:1
Fuel system	Indirect injection
Max. power (DIN)	67bhp at 4,000rpm and 114lb ft at 1,800rpm

Transmission

Permanent four-wheel drive (selectable four-wheel drive optional on 4-cylinder models, 1983–4 only)

Main gearbox

(a) With 4-cylinder engines:
5-speed LT77 manual gearbox
Ratios 3.585:1, 2.300:1, 1.507:1, 1.000:1, 0.830:1, reverse 3.700:1
(b) 1983–4 model years, with V8 engine:
4-speed LT95 manual gearbox
Ratios 4.069:1, 2.448:1, 1.505:1, 1.000:1, reverse 3.664:1
(c) 1986 model year, with V8 engine:
5-speed LT85 manual gearbox
Ratios 3.65:1, 2.18:1, 1.43:1, 1.1:1, 0.79:1, reverse 3.82:1

Transfer gearbox

(a) With 4-cylinder engines:
Two-speed LT230
Ratios 1.667:1 (high) and 3.320:1 (low); Ninety models with 1.41:1 high ratio
(b) 1983–4 model years, with V8 engine:
Two-speed transfer box integral with main gearbox
Ratios 1.336:1 (high) and 3.321:1 (low)
(c) 1986 model year, with V8 engine:
Two-speed LT230
Ratios 1.41:1 (high) and 3.32:1 (low)
Axle ratio: 3.54:1

Suspension and Steering

Suspension

Front	Radius arms and Panhard rod
Rear	Radius arms, support rods and central wishbone assembly

Coil springs all round

Continued overleaf

Continued from previous page

SPECIFICATIONS

NINETY, ONE TEN and ONE TWO SEVEN models 1983–86

Suspension and Steering

***Suspension* (continued)**

Boge Hydromat self-energizing rear ride-levelling strut standard on County models and optional on others; rear anti-roll bar standard with levelled suspension from 1986 model year

Steering

Worm and roller steering with 20.2:1 ratio; optional power assistance with 19.3:1 ratio

Brakes

Type	Disc brakes on the front wheels and drum brakes at the rear; dual hydraulic circuits; servo assistance; separate drum-type transmission parking brake

Electrical System

12V with alternator

Dimensions

Ninety models

Track	58.5in (1,486mm)
Wheelbase	92.9in (2,360mm)
Overall length	146.5in (3,720mm)
Overall width	70.5in (1,791mm)
Unladen height	77.3in (1,963mm) for truck cab 77.4in (1,966mm) for soft-top 77.6in (1,971mm) for hardtop and Station Wagon
Unladen weights	3,540lb (1,606kg) minimum with 4-cylinder petrol engine 3,622lb (1,643kg) minimum with diesel engine

One Ten models

Track	58.5in (1,486mm)
Wheelbase	110in (2,794mm)
Overall length	175in (4,445mm) for truck cab and soft-top 180.3in (4,580mm) for Station Wagon 184in (4,674mm) for HCPU
Overall width	70.5in (1,791mm)
Unladen height	80.1in (2,034mm)
Unladen weights	3,743lb (1,698kg) minimum with V8 petrol engine 3,799lb (1,723kg) minimum with 4-cylinder petrol engine 3,840lb (1,742kg) minimum with diesel engine

One Two Seven models

Track	58.5in (1,486mm)
Wheelbase	127in (3,226mm)
Overall length	198in (5,029mm)
Overall width	70.5in (1,791mm)
Unladen height	80.1in (2,034mm)
Unladen weight	4,167lb (1,890kg) minimum with V8 petrol engine 4,222lb (1,915kg) minimum with 4-cylinder petrol engine 4,244lb (1,925kg) minimum with diesel engine

Performance

Ninety Station Wagon with 114bhp V8 and 5-speed LT85

0–60mph	14.2sec
Maximum	83.5mph (134km/h)
Fuel consumption	14–16 mpg (20.2–17.7ltr/100km)

Ninety Station Wagon with Diesel Turbo

0–60mph	22.3sec
Maximum	76mph (122km/h)
Fuel consumption	16–23mpg (17.7–12.3ltr/100km)

One Ten Station Wagon with 114bhp V8 and 4-speed LT95

0–60mph	14.2sec
Maximum	79mph (127km/h)
Fuel consumption	13–15mpg (21.8–18.9ltr/100km)

One Ten Station Wagon with 134bhp V8 and 5-speed LT85

0–60mph	15sec
Maximum	84mph (135km/h)
Fuel consumption	13–15mpg (21.8–18.9ltr/100km)

CHAPTER FIVE

GOING EUROPEAN, 1987–1990

One Ten, One Two Seven And Ninety, 1987–1990

As Land Rover reoriented itself towards European markets, it became obvious that the company had to overhaul its product range. Early One Tens had been designed with tough African conditions and military users in mind, but potential buyers in Europe were going to want more creature comforts. They would be less tolerant of the rough and ready features that had been a Land Rover characteristic for so long.

So the mid-1980s saw a team of Solihull's designers and engineers working on a project to civilize the One Ten family. This was known as Project Capricorn and its main aim was to explore the possibilities. So there was no Capricorn vehicle as such, but a number of the refinements first proposed under Capricorn found their way into production in the later 1980s. Others were delayed. For example, the Capricorn team looked at deleting the centre bulkhead from Ninety Station Wagons as a way of allowing the seats to move back further and so create more leg room. This idea did not enter production at Solihull until 2001, on the 2002 models, although it was adopted earlier on Defender 90s built in South Africa and on the special edition 90SV. Forward-facing rear seats in a Ninety were another Capricorn proposal and those did not make production (by then in rather different form) until the 2007 models.

Even so, the first fruits of the Capricorn project soon became visible and there was a very noticeable step-change for Land Rovers as the 1987 model year opened. It introduced a period in which the Station Wagon models were gradually developed into more credible alternatives to conventional family estate cars. Not only did this make the Ninety and One Ten more attractive to European customers; it also helped to prepare the ground for Land Rover's new dedicated family 4×4, still under development at this stage as Project Jay, but to become the Discovery on its announcement in autumn 1989. From 1987, there was also a new focus on improving road performance and a major element in achieving that was the arrival of a turbocharged diesel engine.

Some of Land Rover's overseas branches had been demanding a more powerful diesel for many years. Australia and South Africa had broken ranks and had developed their own Land Rover derivatives using bought-in engines in the early 1980s; Santana in Spain had developed their own 6-cylinder diesel as early as 1977, but it had never appeared in a coil-sprung model. However, Land Rover at Solihull had never been persuaded that a more powerful diesel engine was necessary until the collapse of the marque's African markets in the early 1980s had brought about a new focus on first-world exports and in particular exports to Europe. On the European continent, Japanese 4×4 makers were doing well with powerful diesel models and it was clear that Land Rover had to do the same if it was to compete successfully in that market.

The Product Planning Department came up with a two-pronged strategy. The medium-term plan was to develop a new version of the existing 2.5-litre diesel engine with a direct-injection system and a turbocharger. This was known as Project Gemini. It would take three to four years, so there was also a short-term plan which depended on adding a turbocharger to the existing indirect-injection engine. This was known as Project Falcon and it was the Falcon engine that was introduced in October 1986.

This was far from the first time the 4-cylinder diesel had been turbocharged. Land Rover had experimented with a turbocharged derivative in the early 1960s, but this had not gone into production. Santana in Spain had put a turbocharged derivative of the 2.25-litre engine into production in the early

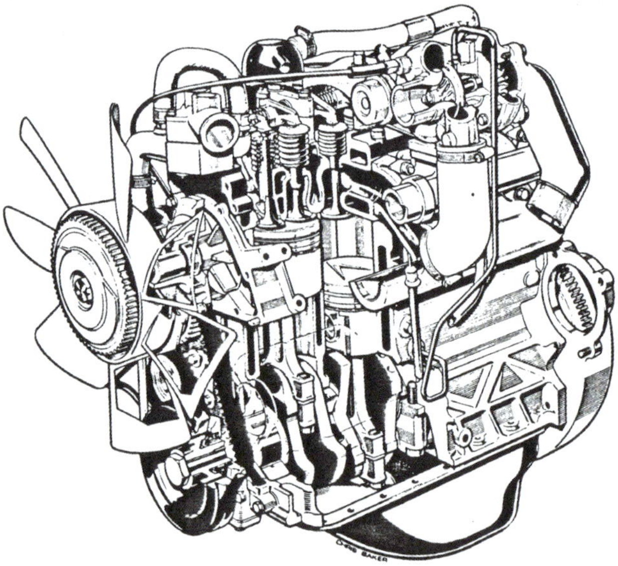

The new Diesel Turbo engine for 1987 was an evolution of the 2.5-litre naturally aspirated diesel engine. The turbocharger was mounted high up, for good accessibility.

1980s. Solihull's engineers were able to draw on both projects for the Falcon engine and may even have looked at some of the aftermarket conversions that had sprung up in the early 1980s as well.

There was more to the job than simply bolting on a turbocharger. As the project progressed, Engineering Director Bill Morris ensured that there was cross-fertilization between the Falcon and Gemini projects, largely to minimize costly and unnecessary differences between the two. So the Falcon cylinder block already incorporated some of the changes that would be needed for the Gemini engine when it appeared three years later. The Falcon engine was never going to give the huge power and torque increases that were available with direct injection, but it did deliver 25 per cent more power (85bhp at 4,000rpm) and 28 per cent more torque (150lb ft at 1,800rpm) than the naturally aspirated 2.5-litre diesel. That translated at around 5sec off the 0–50mph (0–80km/h) time and about 6mph (9.7km/h) more on the top speed. By traditional Land Rover standards, those performance increases were huge.

The 1987 Model Year

That turbocharged diesel engine was the most important new feature when the 1987 models were revealed to the public at the UK Motor Show on 9 October 1986. It was marketed as the Diesel Turbo to distinguish it from the bought-in engine introduced in April 1986 for the Range Rover Turbo D and was an addition to the three existing engine options.

The new 'road performance' orientation for the One Ten family was also apparent from a change to the V8 petrol engine. European and UK Land Rovers now followed those for other markets in having the more powerful 134bhp version of the V8 engine with twin SU carburettors instead of the earlier Zenith-Strombergs.

Several new features distinguished the 1987 models from the outside, too. Most obvious were the new door handles, with car-like plastic grips and push-button releases instead of the traditional Land Rover inset handle. The doors could now be slam-locked (with a double action for the front doors and a single action for the rears on Station Wagons) and there were now two keys – one for the ignition and one for the other locks. County models took on new side stripes with a 'hockey-stick' design and their wheel-arch eyebrows were now painted to match the body. All models gained a styled locking fuel filler cap in black and Diesel

LEFT: **There were new County stripes for the 1987 models, with a 'hockey-stick' design. There were different stripe colours to suit different body colours.**

BELOW: **The first year's production of Diesel Turbo models had tail decals in italicized capitals.**

Turbo models had a large Turbo decal in capital letters on the tail panel.

Less obvious was a switch to aluminium cantrails (from VIN 260482), which were now part of the roof assembly and were painted to match it. New roof rack designs became necessary because racks intended for the old steel rails were likely to damage the new design. On Ninety models, Rostyle styled steel wheels (originally designed for the Range Rover) became an option, always finished in Ivory White. On top of that, the 1987 models were slightly less prone to water leaks into the body than earlier types; a lot of work had been done on this in order to improve overall refinement.

On the inside, the door trims now had a new release handle and flush-fitting sill-locking buttons. There were several changes to the seats, too. The nine-seater and ten-seater Station Wagons gained an asymmetrically split back seat (although the twelve-seaters still had a one-piece bench), while the inward-facing individual rear seats were relocated to give more shoulder room. Front seats now had inertia-reel belts with smaller reels and One Ten Station Wagons had inertia-reel belts in the rear. Then, shortly after the start of the new model year in November 1986, the fore-and-aft adjustment was deleted from the front passenger seat and County front seats lost their moulded back panels in favour of leathercloth coverings.

Symbolic of the move towards a more car-like ambience was the arrival of an interior bonnet release, while windscreens were now laminated, in either plain or tinted glass. A choke warning light was added and a single twin-jet windscreen washer replaced the earlier pair of single-jet types. On Station Wagons, the rear washer was now above the door, while a single reservoir under the bonnet now served both front and rear systems, replacing the earlier seperate reservoirs.

Legislation was behind a change in the lighting arrangements. The Road Vehicle Light Regulations (1984) came into effect and Land Rover met them with what the company called dim-dip lighting. With the engine off, the side-light position on the switch gave side lights only for parking, but with the engine running, that same position gave reduced-output dipped beam headlamps. However, very early 1987 models were built without this feature.

There were two more minor changes during the season before a further set of major changes in the summer. Beginning in February 1987 on the production lines, two extra black plastic air vents were added above the tail door of hardtops and Station Wagons to improve ventilation. In April, Venetian Red was deleted from the colour range, leaving just six colour choices; Land Rover dealers were promised new colours for later in 1987.

The next group of changes occurred in July 1987, but they were seen as mid-season changes because the 1988 model year was not scheduled to begin until the end of the year. In the UK, several low-volume models were deleted: soft-top Ninety and One Ten models disappeared; so did all the V8 derivatives of the Ninety; and One Ten V8 hardtops and High-Capacity Pick-Ups were deleted from the catalogues. These models remained available for other markets, of course.

The July 1987 changes also brought County vinyl seats as standard for non-County models, which in effect meant the standard seats now enjoyed the same range of adjustment as the earlier County type. These were accompanied by head restraints on the outer pair, except on pick-ups. Standardized right across the range were a heater-demister and floor

This publicity picture of a 1987 model Ninety County shows the optional Rostyle wheels introduced that year; 1987 was also the last year for the galvanized bumper finish.

This still from the TV advertisement shows the Ninety winching itself up the face of a dam in Wales.

and roof trims. All models gained a fuel-filler tube extension and an under-ride protection bar. Side and rear folding steps were standardized on Station Wagons and all diesels now came with an oil cooler. On One Ten Station Wagons, the rear suspension levelling system was deleted from the standard specification and became an extra-cost option.

Particularly memorable in the 1987 model year was a TV advertisement for the Ninety and One Ten. It featured a Ninety winching itself up an enormous dam, to the accompaniment (of course) of the 'Dambusters' March'. The vehicle was registered D415 BDU and the dam was at Llyn Clywedog in Wales. Despite scepticism in some quarters, the whole event really did happen the way it appeared to on screen.

Before its Time – The Cariba Concept Vehicle

As Land Rover worked to focus more effectively on European markets in the mid-1980s, ideas came thick and fast from all over the company. One idea that went public came from the Public Relations Department rather than from Engineering and was called the Cariba. It was announced in autumn 1987 and was clearly inspired by the new thinking that was going into Project Capricorn.

The Cariba showed a different side of the Land Rover's character – one that was already being exploited by enthusiasts and aftermarket specialists. It was designed as a 4×4 fun

The Cariba was not a mainstream Land Rover project, but it picked up on possibilities for developing the Ninety and anticipated the special North American derivative by several years.

STANDARD COLOURS AND TRIMS — 1987 Model Year

Paints

Bronze Green	LRC001
Stratos Blue	LRC327
Ivory White	LRC354
Trident Green	LRC325
Marine Blue	LRC006
Venetian Red *	LRC440
Slate Grey	LRC348

All colours were available on County models.
* Venetian Red was deleted in April 1987.

Roofs, Wheels and Soft-Tops
Roofs and wheels were always painted Ivory White. Soft-tops were Matt Stone.

Side Stripes
County models had 'hockey stick' side stripes in two-tone Blue, Green or Grey, with the County name above the lower section of the stripe on each front door and a V8 identifier where appropriate.

Upholstery and Trim
Standard upholstery was Black vinyl. County models had County Brown cloth upholstery.

vehicle rather than a workhorse, taking the Ninety into the emerging 'leisure' market. One problem would have been the high cost as compared to similar models from Japanese makers (notably the Suzuki SJ), but the real reason why the idea went no further was because it emerged at a time when Solihull had committed to entering the leisure market in a different way, through the family-oriented vehicle that would be announced as the Discovery in 1989.

The Public Relations Department had the vehicle converted at Solihull from a redundant Ninety V8 press demonstrator, registered D347 FDU. It was created with minimum difficulty by using items that were for the most part already available in the Land Rover range. The basic concept was an open vehicle with the performance of the V8 engine and an eye-catching appearance.

A unique metallic light blue based on two contemporary Range Rover options helped the visual appeal. The Cariba had wide tyres on aftermarket spoked wheels, a rollover bar braced to the windscreen header rail and frameless door windows. A military-specification swing-away spare wheel carrier was mounted to the side-opening tailgate, a bull bar to the front and there was a custom-made soft-top over the cab and a tonneau cover for the back body. Inside, the Cariba had specially trimmed Range Rover leather front seats, with rear seats retrimmed to match.

The Cariba was very well received by those members of the press who drove it, but it quickly disappeared from view. Senior management at Solihull saw it as a maverick project that was not representative of the company's future direction and the vehicle was sold to the public through a Land Rover dealer in Cheltenham, supposedly ending up in either Greece or Cyprus.

The 1988 Model Year

The 1988 model year was an eventful one that witnessed major changes for Land Rover, both planned and unplanned. Key among the planned events were celebrations focused on the fortieth anniversary of the Land Rover marque in April 1988, along with changes to the company's logos and branding identities. There was also a change of ownership, as the British Government divested itself of responsibility for the Rover Group and passed it on to British Aerospace.

The unplanned event was a major strike on the Solihull site that began on 23 February, when about 6,000 workers walked out in a dispute over pay. The strike was called off on 28 March, but it did disrupt production long enough to affect the vehicle allocations to dealers for June. It also obliged Land Rover to call off the fortieth anniversary open days at Solihull that had been planned for 30 April and 1 May. A further casualty was a plan to build forty Ninety models and register them all with number plates including the number 40. In the event, only two were built.

For 1988, a proper radio installation in the centre of the dashboard was available and the ashtray was displaced to the top rail of the dash. This is a County Station Wagon, with leather-bound steering wheel.

The Ninety truck cab remained a popular derivative. Black bumpers became standard on 1988 models.

One was registered E40 KDU and was built to a specification that was not standard in the UK, in order to represent the 'traditional' short-wheelbase models. So it had a soft-top (no longer available on Nineties for the UK) that was finished in military Khaki instead of the then-current Matt Stone and was painted Bronze Green with the wheels in Bronze Green instead of the standard production Ivory White. It also had its wheel-arch eyebrows painted to match the body. This vehicle still survives on Land Rover's own historic vehicles fleet. The second vehicle was registered as F40 FWK, which appeared in public in 1989 after being fitted with floats for use on the water at Cowes Week during Land Rover's sponsorship of the event.

Production of the 1988-model Land Rovers began in November 1987 and they were announced to the public at the Royal Smithfield Show that opened on 6 December. There were some major exterior changes, as the front bumper and rear grab handles switched from a galvanized finish to black. Though still galvanized as a first stage, they were powder-coated as well and the rear vertical edges of the bumper were finished to improve its appearance.

There were V8 decal badges for V8-engined models, while Diesel Turbo models had a new rear decal reading 'Turbo' in lower-case letters instead of the earlier italicized capitals.

For all Station Wagons, a tilt-and-remove glass sun-hatch option was made available. For the County Station Wagons only, grille and headlamp surrounds were now painted to match the body, giving an appearance that was intended to be more car-like but actually seemed rather strange at the time. Other models retained their black grilles and headlamp panels.

The headlamp power wash that had been available on Station Wagons and standard on County derivatives now disappeared altogether. There were new side decals for the County models, too, in graduated grey with an angled section on each front door.

Interior changes were immediately obvious on opening a door. Although there was still vinyl upholstery on base models and cloth with the County specification, all trim was now grey. Front seats on all models had both a reclining squab and fore-and-aft adjustment (and were described as County-spec types). On the One Ten Station Wagons, a neat one-piece moulded rear side-trim panel replaced the five separate pieces on earlier models and there was a matching B-post trim finisher as well. The steering wheel remained black right across

E40 KDU was one of the planned forty commemorative vehicles to be built for Land Rover's fortieth anniversary. A strike at the Solihull plant ensured that only two were actually completed.

STANDARD COLOURS AND TRIMS

1988 Model Year

Paints

Arrow Red	LRC390
Shire Blue	LRC392
Bronze Green	LRC001
Slate Grey	LRC348
Ivory White	LRC354
Trident Green *	LRC325

All colours were available on County models.
* Trident Green was not available after July 1988.

Roofs, Wheels and Soft-Tops

Roofs and wheels were always painted Ivory White. Soft-tops were Matt Stone.

Side Stripes

County models had graduated Grey side stripes, with the County name in angled sections on each front door.

Upholstery and Trim

Standard upholstery was Grey vinyl. County models had Grey cloth upholstery.

the range, although there was a new centre finisher, and on County Station Wagons the rim of the wheel was bound in leather. Up above, the interior light had moved rearwards to make room for the optional sunroof and when that was fitted an optional Sun-hatch Accessory Pack could also be had, consisting of a wind deflector, a storage bag for the glass roof and a clip-on interior roof panel.

Changes in fuel duty introduced under the 1988 Budget in March favoured unleaded petrol, so in April Land Rover contacted its dealers with confirmation that all current production models could safely use unleaded fuel. For the V8 petrol engines, however, the company recommended retarding the timing by 3 degrees; for those customers running older 2.3-litre petrol engines, the recommendation was to fit the latest cylinder head from a 2.5-litre engine, or to fit the hardened exhaust valve inserts used in that engine.

There were then new UK seating regulations from May for vehicles classified as Dual Purpose types and these reduced the requirement from two front passenger seats to one. So instead of the centre front seat, Station Wagons could be specified with a cubby box at no extra cost, reducing the seating capacity from ten to nine. There were probably few takers, though, because these vehicles remained subject to the tax that encouraged most UK buyers to buy the tax-exempt twelve-seater models. To retain their seating capacity and tax exemption, these models were not available with a centre cubby box.

There was one further running change during the season, as Trident Green was deleted from the paint options in July 1988. Land Rover promised its dealers new colours some time later in the calendar year. In practice, that meant December 1988, when the 1989 models were announced.

The Australian 110 Dual Cab

Jaguar Rover Australia built a fleet of twenty-five 'specials' to tow exhibition trailers during the Australian Bicentennial celebrations in 1988. These were based on One Ten V8 Station Wagons, cut down to create five-seater pick-ups that the Australians referred to as Dual Cab utes (utilities). A fifth wheel was installed in the back body to allow the trailer to be towed on a swan neck coupling.

After the celebrations were over, the Australian Bicentennial Authority sold these vehicles off to the public, and several still survive.

The 1989 Model Year

The Royal Smithfield Show that opened on 4 December 1988 was where Land Rover announced its 1989 utility models for the UK. As in 1987, this show had been chosen as a way of ensuring that the publicity focus could be on the Range Rover at the major motor shows in the autumn. Actual production of the 1989 models began in November 1988.

The focus was once again on making the Station Wagons more acceptable to car users, both to increase sales generally

The Australian Dual Cab ute (utility vehicle) anticipated the Defender Double Cab by around thirteen years, although it was constructed differently. Many were white with blue, rather than the blue with white of this example.

and to promote the idea that Land Rovers could be everyday vehicles. This latter was important because the new Discovery family model was scheduled for release in autumn 1989.

Several low-volume derivatives disappeared from the UK lists. Most of these were normally only supplied to fleet users and those which were axed were the Ninety naturally aspirated diesel truck cab, all 4-cylinder petrol and naturally aspirated diesel One Ten truck cabs and the naturally aspirated diesel derivative of the One Ten Hi-Cap. Although V8 models were still listed, they were in practice almost non-existent in the UK for 1989.

The key changes had been designed for the Station Wagon models, affecting their roof and upper side panels. However, where these panels were also used on hardtop derivatives, the hardtops also benefitted. The biggest change was to a smooth roof panel with no external ribs; these had been replaced by longitudinal stiffeners inside the roof. The upper body side panels had also been redesigned to give a smooth exterior finish without visible rivets or spot welds; a new frame and skin structure made this possible. On Ninety hardtops, a new Roof Appointment Pack option brought the tilt-and-remove sun hatch, Alpine lights and a full-length headlining as well. In the UK, the floor and roof trim options were now standardized. A further addition for all models with a full-length roof was a 'hands-free' automatic check strap for the tail door.

Then there were of course some new colours to brighten the vehicles' appearance. Arran Beige and Eastnor Green did exactly that, but with the arrival of Eastnor Green died a long Land Rover tradition, as the Bronze Green available

The second vehicle built for the fortieth anniversary was declared surplus to requirements and became the Floating Ninety used for publicity at Cowes Week in 1989. The design of the floats was taken from an early 1960s military Land Rover design that had not gone into volume production.

Details of special vehicles built for the Royal Family are not normally released to the public, but this long-wheelbase Station Wagon dating from 1989 was probably based on a 127 chassis. The raised roof was needed to give headroom for the rear seats, which had been moved back over the rear axle.

since 1949 disappeared from the options list. All models, not just County types, now had side stripes with a diagonal section across the front doors. The County stripes were the same as for 1988, but the other stripes were in two-tone grey, with '90' or '110' as appropriate on the door section. The reason for this proliferation of stripes was not apparent at the time, but Land Rover was actually preparing the way for the new Discovery model due the following autumn, which would have side stripes as standard. Buyers who

For 1989, the County stripes had diagonal sections on the front doors. The One Ten has the slide-and-tilt sunroof and the Ninety shows the short-lived air vents above the tail door. The body-colour grille and headlamp surrounds had been new for 1988.

STANDARD COLOURS AND TRIMS — 1989 Model Year

Paints

Arran Beige	LRC433
Ivory White	LRC354
Arrow Red	LRC390
Shire Blue	LRC392
Eastnor Green	LRC419

All colours were available on County models.

Paint names were changed to Beige, Blue, Green, Red and White in August 1989; all colours remained the same except for Blue, which switched to LRC424.

Roofs, Wheels and Soft-Tops
Roofs and wheels were always painted Ivory White. Soft-tops were Matt Stone.

Side Stripes
All models had two-tone Grey side stripes. These had 'County', '90' or '110' as appropriate in angled sections on each front door.

Upholstery and Trim
Standard upholstery was Grey vinyl. County models had Grey cloth upholstery.

objected to the side stripes on their Nineties and One Tens could have them deleted at no extra cost.

There were changes to the interior again, too. The basic front seats (known to Land Rover as 'Fleet only' types) were withdrawn and the front seats on County models were modified to allow the driver to sit 25mm (just over an inch) further back; it was a small change, but every little helped to make the vehicles more comfortable and car-like. A new radio-cassette unit, standard on County models but optional elsewhere, helped as well. The new one was a Philips DC550, which replaced the DC552 and DC556 units used earlier, and added sixteen pre-sets to the radio functions. Air-conditioning remained on the options list, but for 1989 was only available when power-assisted steering was also fitted; Land Rover had presumably discovered that most buyers of one also bought the other option as well.

Lastly, the 1989 season also brought some tyre changes. The new standard fitment for Nineties was the Michelin M&S 205 × 16; 7.50 × 16s were still an option, but the 6.00 × 16 size was now withdrawn. Standard wear for One Tens remained Avon Ranger 7.50 × 16s, but the Pirelli LR04 7.50 × 16s were withdrawn and One Ten V8s and County models now came with Avon Rangemaster 7.50 × 16 radials.

Towards the end of the season, in May 1989, seat belts were standardized for all forward-facing seats on UK models. This meant that the mini-reel inertia belts already used for outer seats were supplemented by lap belts as standard for centre front and (on Station Wagons) centre rear seats as well. Then in August, a new policy on paint colours was announced, as Shire Blue was replaced by a new blue colour. In future, the colour names on all utility Land Rovers would be descriptive rather than evocative. The main reason for this was to allow the new paint names for the forthcoming Discovery to sound more exciting than those used on the County models; in practice, some of them would in fact be exactly the same paints under different names.

The 1990 Model Year

The autumn of 1989 was a very busy time for Land Rover, as the company introduced two new engines for the Range Rover (a 3.9-litre petrol V8 and a 2.5-litre diesel) and the Discovery arrived as a completely new model. The One Ten family, and in particular the Station Wagons, had been carefully groomed over the last three years in order to prepare for what amounted to a major realignment of the Land Rover product range, so for the 1990 model year it was left largely alone. Another important change in this period, announced in June 1990, was that Land Rover severed its relationship with Santana (Metalúrgica de Santa Ana) in Spain.

Santana's relationship with Land Rover had begun in the mid-1950s, when it had begun building leaf-sprung models from CKD kits. Written into the agreement between the two companies had been that Santana would be allowed to progress towards full local manufacture, which it had achieved in 1967. Subsequently, Santana had developed its Land Rovers further to cater for its own domestic market and for requirements in its overseas markets, which were mostly in Central and South America. However, Santana had not adopted coil springs in the early 1980s, so the vehicles it was building in 1990 were still leaf-sprung – although they had advanced considerably from the original designs.

During the 1980s, there had been a close and valuable engineering relationship between Solihull and Santana, with the Spanish company designing the original wind-up door windows introduced on UK assembly in 1984 and also the LT85 5-speed gearbox used with V8 engines after 1985. However, Land Rover decided that it was time for the relationship to end because the Spanish models, which carried the Land Rover name in addition to Santana's, were no longer representative of the current Land Rover product. There were Santana models after 1990, and they looked like Land Rovers, but they were built entirely in Spain and used Santana's own versions of some Land Rover components.

There was one distinguishing novelty for the 1990-model Land Rovers, despite Solihull's focus on its other products. Instead of grille badges reading 'Land Rover 90' or 'Land Rover

ABOVE: **For 1990, the grille badges changed to an all-number type and a green Land Rover oval badge was added to the grille slats. These are seen on a One Ten used by a fire brigade, which explains the all-red colour and the blue light.**

ABOVE: **The 1990 models all had side decals, unless deleted to special order. This Ninety hardtop also shows the lower-case lettering used on Diesel Turbo models from 1988 onwards.**

LEFT: **Decals mattered by the end of the 1980s. The ones on this Ninety Diesel Turbo Station Wagon were an option in France. Despite Land Rover's careful distinction between Range Rover Turbo D and Land Rover Diesel Turbo, the French decals call this Ninety a Turbo D!**

STANDARD COLOURS AND TRIMS

1990 Model Year

Paints

Beige	LRC433
Red	LRC390
Blue	LRC424
White	LRC354
Green	LRC419

All colours were available on County models.

Roofs, Wheels and Soft-Tops

Roofs and wheels were always painted Ivory White. Soft-tops were Matt Stone.

Side Stripes

All models had two-tone Grey side stripes. These had 'County', '90' or '110' as appropriate in angled sections on each front door.

Upholstery and Trim

Standard upholstery was Black vinyl. County models had County Brown cloth upholstery.

CAMEL TROPHY VEHICLES

One Ten Station Wagons were the team vehicles on the 1988 and 1989 Camel Trophy events and, as before, there were One Ten and One Two Seven support vehicles on all the events of the late 1980s.

On both occasions, the team vehicles were One Ten Diesel Turbo Station Wagons. The 1988 event was held in Sulawesi, with twelve international teams, and the 1989 event was held in the Amazon Basin, with fourteen international teams.

The 1989 Camel Trophy was won by the British team, brothers Bob and Joe Ives.

110', they had new badges which simply read '90', '110'; even the One Two Seven lost its old printed metal Land Rover badge and took on a new one, matching the others, that read '127'.

From December 1989, the 127 also had a higher payload capacity as a consequence of the increase to a 3,500kg (7,716lb) gross vehicle weight (GVW). UK dealers were also reminded that there was no need to have 127 models plated and tested locally to the new standards because the model was already fully homologated. There was one other change during the 1990 model year, in May 1990, when the rear roof vents were deleted from hardtop and Station Wagon bodies. Though they had seemed like a good idea on their introduction in early 1987, when they had been intended to improve ventilation, in practice they allowed water and dust to get into the bodies.

VINS

The VIN codes used in this period consisted of seventeen characters, made up of an eleven-character prefix code and a six-digit serial number.

Example: SALLDHAB1BA-xxxxxx

This breaks down as follows:

SAL	Manufacturer code (Austin Rover to 1986; Rover Group thereafter)
LD	Land Rover coil-sprung utility
H	110in (2,794mm) wheelbase
	K = 127in (3,226mm) wheelbase
	M = Australian Heavy-Duty 6×6
	S = 110in (2,794mm) military specification
	V = 92.9in (2,360mm) wheelbase (Ninety)
A	Truck cab, soft-top or hardtop body
	E = Two-door crew cab (export only)
	F = Four-door crew cab
	H = HCPU
	M = Station Wagon
B	2.5-litre Diesel Turbo engine
	C = 2.5-litre diesel engine
	D = 2.5-litre petrol engine
	G = 2.3-litre diesel engine
	H = 2.3-litre diesel engine
	V = 3.5-litre petrol engine
	Z = 3.9-litre Isuzu diesel engine (Australia only)
1	RHD with 4-speed manual gearbox
	2 = LHD with 4-speed manual gearbox
	7 = RHD with 5-speed manual gearbox
	8 = LHD with 5-speed manual gearbox
B	Model years 1985–7
	A = 1983–4
	E = 1988
	F = 1989
	G = 1990
A	Assembled at Solihull
	F = Shipped as KD for overseas assembly
	Y = Part-KD, part-Australian built (Heavy-Duty 6x6 only)
xxxxxx	Serial number

Engine Identification Numbers

Engines have a nine-digit identification number. The digits break down into three groups: the first group of three digits identifies the engine type, the second group of five digits is the engine serial number and the final digit identifies su-groups created by running specification changes.

Example: 10H-00001A.

2.3-litre petrol engine

10H-00001A	8:1 CR, non-detoxed
11H-00001A	8:1 CR, detoxed
13H-00001A	7:1 CR

2.3-litre diesel engine

10J-00001A	Standard engines

2.5-litre diesel engine

12J-00001A	Standard engines
18J-00001A	Military engines with 24V electrical system

2.5-litre petrol engine

17H-00001A	Standard engines
18H-00001A	Military engines with 24V electrical system
19H-00001A	For auto gearbox (unsuccessful Swiss military tender)

2.5-litre Diesel Turbo engine

19J-00001A	Standard engines

3.5-litre V8 carburettor engine

14G-00001A	8.13:1 CR, non-detoxed, with LT95 gearbox
15G-00001A	8.13:1 CR, detoxed, with LT95 gearbox
18G-00001A	Military engines with 24V electrical system and LT85 gearbox
19G-00001A	8.13:1 CR, for Saudi Arabia, with LT95 gearbox
20G-00001A	8.13:1 CR, non-detoxed, with LT85 gearbox
21G-00001A	8.13:1 CR, detoxed, with LT85 gearbox
22G-00001A	9.35:1 CR, for Australia
23G-00001A	8.13:1 CR, detoxed, for Llama forward-control
24G-00001A	8.13:1 CR, detoxed

VINS AND ENGINE NUMBERS 1983–1990 Models

3.5-litre V8 injection engine

31G-00001A 8.13:1 CR, for auto gearbox (unsuccessful Swiss military tender)

Isuzu 4BD1 3.9-litre diesel

Isuzu 4BD1T 3.9-litre turbocharged diesel

The Isuzu engines use a completely different identification system to the Land Rover one. Isuzu engines of all types in production at any one time were numbered in the same serial numbering sequence, beginning at 100000 and ending with 999999. Once the final number had been reached, numbering recommenced at 100000.

Individual engines can only be identified by the use of this serial number in conjunction with a production listing which determines the type and production date. However, it is possible to establish the age of an engine from a date-code cast into the block.

SPECIFICATIONS NINETY, ONE TEN and ONE TWO SEVEN models 1987–90

Petrol engines

Type	2.5-litre
Bore and stroke	90.47mm × 97mm
Capacity	2,495cc
Valves	4-cylinder with ohv
Compression ratio	8:1
Carburettor	Single Weber twin-choke carburettor
Max. power (DIN)	83bhp at 4,000rpm and 133lb ft at 2,000rpm

Type	V8
Bore and stroke	88.9mm × 71.1mm
Capacity	3,528cc
Valves	V8-cylinder with pushrod ohv
Compression ratio	8.13:1
Carburettor	Two SU carburettors
Max. power (DIN)	134bhp at 5,000rpm and 187lb ft at 2,500rpm

Diesel engines

Type	2.5-litre
Bore and stroke	90.47mm × 97mm
Capacity	2,495cc
Valves	4-cylinder with ohv
Compression ratio	21:1
Fuel system	Indirect-injection
Max. power (DIN)	67bhp at 4,000rpm and 114lb ft at 1,800rpm

Type	Diesel Turbo
Bore and stroke	90.47mm × 97mm
Capacity	2,495cc
Valve	4-cylinder with ohv
Compression ratio	21:1
Fuel system	Indirect-injection and Garrett T2 turbocharger
Max. power (DIN)	85bhp at 4,000rpm and 150lb ft at 1,800rpm

Transmission

Permanent four-wheel drive

Main gearbox

(a) With 4-cylinder engines:

5-speed LT77 manual gearbox

Ratios: 3.585:1, 2.300:1, 1.507:1, 1.000:1, 0.830:1, reverse 3.700:1

(b) With V8 engine:

5-speed LT85 manual gearbox

Ratios: 3.65:1, 2.18:1, 1.43:1, 1.1:1, 0.79:1, reverse 3.82:1

Transfer gearbox

(a) With 4-cylinder engines:

Two-speed LT230

Ratios 1.667:1 (high) and 3.320:1 (low); Ninety models with 1.41:1 high ratio

Continued overleaf

Continued from previous page

SPECIFICATIONS

NINETY, ONE TEN and ONE TWO SEVEN models 1987–90

(b) With V8 engine:
Two-speed LT230
Ratios 1.41:1 (high) and 3.32:1 (low)
Axle ratio: 3.54:1

Suspension and Steering

Suspension

Front	Radius arms and Panhard rod
Rear	Radius arms, support rods and central wishbone assembly

Coil springs all round
Boge Hydromat self-energizing rear ride-levelling strut standard on County models and optional on others; rear anti-roll bar standard with levelled suspension from 1986 model year

Steering

Worm and roller steering with 20.2:1 ratio; optional power assistance with 19.3:1 ratio

Brakes

Type Disc brakes on the front wheels and drum brakes at the rear; dual hydraulic circuits; servo assistance; separate drum-type transmission parking brake

Electrical System

12V with alternator

Dimensions

Ninety models

Track	58.5in (1,486mm)
Wheelbase	92.9in (2,360mm)
Overall length	146.5in (3,720mm)
Overall width	70.5in (1,791mm)
Unladen height	77.3in (1,963mm) for truck cab 77.4in (1,966mm) for soft-top 77.6in (1,971mm) for hardtop and Station Wagon
Unladen weights	3,540lb (1,606kg) minimum with 4-cylinder petrol engine 3,622lb (1,643kg) minimum with diesel engine

One Ten models

Track	58.5in (1,486mm)
Wheelbase	110in (2,794mm)
Overall length	175in (4,445mm) for truck cab and soft-top 180.3in (4,580mm) for Station Wagon 184in (4,674mm) for HCPU
Overall width	70.5in (1,791mm)
Unladen height	80.1in (2,034mm)
Unladen weights	3,743lb (1,698kg) minimum with V8 petrol engine 3,799lb (1,723kg) minimum with 4-cylinder petrol engine 3,840lb (1,742kg) minimum with diesel engine

One Two Seven models

Track	58.5in (1,486mm)
Wheelbase	127in (3,226mm)
Overall length	198in (5,029mm)
Overall width	70.5in (1,791mm)
Unladen height	80.1in (2,034mm)
Unladen weights	4,167lb (1,890kg) minimum with V8 petrol engine 4,222lb (1,915kg) minimum with 4-cylinder petrol engine 4,244lb (1,925kg) minimum with diesel engine

Performance

Ninety Station Wagon with Diesel Turbo

0–60mph	22.3sec
Maximum	76mph (122km/h)
Fuel consumption	16–23mpg (17.7–12.3ltr/100km)

One Ten Station Wagon with 134bhp V8 and 5-speed LT85

0–60mph	15sec
Maximum	84mph (135km/h)
Fuel consumption	13–15mpg (21.8–18.9ltr/100km)

CHAPTER SIX

THE FIRST DEFENDERS, 1991–1994

The 200Tdi Period, 1991–4 Model Years

The 1991 model year marked a major watershed in the development of the coil-sprung Land Rovers for two main reasons. The first was that what had formerly been simply a Land Rover became a Land Rover Defender. The second was that, for the first time ever, the Land Rover utilities built at Solihull could boast a world-class diesel engine.

So why should Land Rover suddenly decide to call its workhorse model by a new name? The thinking within the company was that it was now time to distinguish the marque name from the product. The real catalyst for change had been the arrival of the Discovery as a third model range during 1989; before then, there had been Land Rovers and Range Rovers, but now the description 'a Land Rover' could apply just as well to the Discovery as to the company's traditional product.

The new name was dreamed up by a group of managers from Land Rover UK and Range Rover of North America during a get-together in a bar in downtown Boston in the USA. It fitted well, carrying on the idea of model names beginning with D which had been initiated with the Discovery (although there would actually be no more afterwards) and representing a rather politically correct view of the Land Rover's role in military forces around the world. Though it jarred a little at first on those who had long associations with the Land Rover marque, it soon became accepted. Land Rover themselves susequently used it to describe earlier coil-sprung models as well, which ultimately led to a certain amount of confusion.

What became a Land Rover Defender in autumn 1990 had never been intended as a totally new model. It was simply the latest iteration of the coil-sprung Land Rover, incorporating a number of evolutionary improvements which included a new engine – the 200Tdi turbocharged diesel that had made its first appearance in the Discovery a year earlier. This new version of the coil-sprung Land Rover was developed under an engineering programme known as Project Titan – although several documents from Solihull's engineering department call it Tyton; obviously, spelling was not every engineer's strong point!

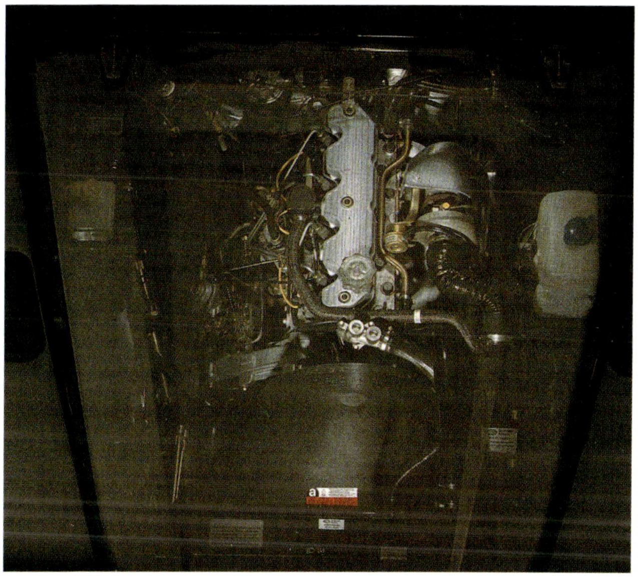

The Gemini, or 200Tdi engine, seen here as a 'ghost' image through a transparent bonnet, was a further evolution of the long-serving 4-cylinder diesel, despite its display of modern technology.

The story of the new diesel engine's development is an interesting one. Land Rover was only too aware that it needed a credible diesel engine for its workhorse models, so as early as 1985 the company had an engineering strategy in

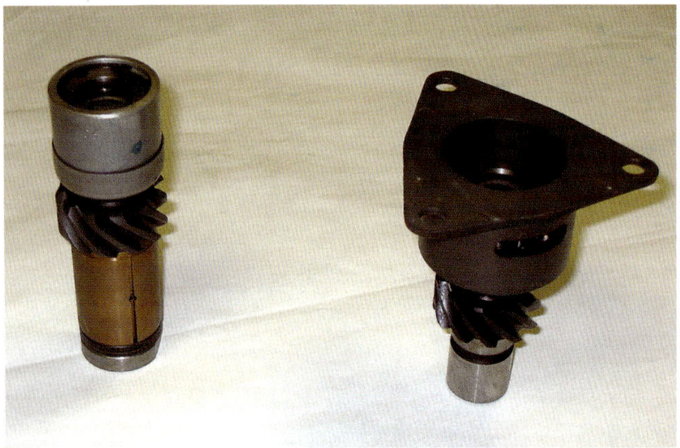

One development problem encountered late in the Gemini programme was the skew gear that drove the diesel pump. The existing type (*left*) was not strong enough for the new application and had to be redesigned (*right*).

place. With the assistance of the Product Planners, notably John Bilton who was in charge of power-train strategies for Land Rover, they formed an alliance with AVL, an Austrian company that specialized in diesel engine development. The Austrians at the time were working on adapting direct-injection technology to small diesel engines; the end result was expected to be quite spectacular improvements in combustion efficiency, which would give both better fuel economy and greater power and torque.

Direct-injection, where the air–fuel mixture is injected directly into the combustion chamber, was nothing new in itself. It had been around for many years in diesel engines for heavy commercial vehicles, but had not been used in diesel car engines because it was unacceptably noisy. Car diesels used indirect-injection systems, where the mixture was injected into a pre-chamber before reaching the combustion chamber. This resulted in less efficient combustion, but also considerably reduced combustion noise. All earlier Land Rover diesel engines had been indirect-injection types.

The plan drawn up between Land Rover and AVL was to fit the British company's existing 2.5-litre 4-cylinder diesel engine with a direct-injection cylinder head. There were thoughts of creating both naturally aspirated and turbocharged derivatives of the new engine, which was known as Project Gemini. The Gemini engine was intended purely for the workhorse Land Rovers – although Solihull would obviously have considered selling it to other companies as well, in the way it had done with earlier diesels. AVL was given a diesel Ninety as a development vehicle in 1985 and as the project progressed, there was of course close liaison with the Solihull diesel engineers.

In the meantime, Solihull could not wait until the new direct-injection engine was ready before improving the performance of their diesel Land Rovers. So in a separate project, the company developed a turbocharger installation for the existing indirect-injection engine, which became the Falcon or Diesel Turbo engine that entered production in mid-1986. As progress reports from Austria suggested that modifications to the cylinder block would be beneficial, so a number of these were put through the system early as 'essential' requirements for the Falcon engine. It all helped to spread the cost of different projects within the Engineering budget at Solihull.

The critical stage in the project was reached when AVL fitted the engine with new two-stage diesel injectors, developed by Bosch in Germany. These injected a small pilot charge of fuel mixture to start the combustion process, followed by the rest of the charge to complete it. This took the harsh edge off the characteristic diesel combustion 'crack'. With noise levels reduced dramatically, as well as quite exceptional results from the direct-injection technology in torque, power and fuel consumption, it was clear that the Gemini engine was a winner. The idea of developing a naturally aspirated derivative – and a petrol version as well, which might have replaced the existing 2.5-litre Land Rover petrol engine – was abandoned and work went full steam ahead on the turbocharged diesel, with an air-to-air intercooler to improve performance still further.

Scheduled for 1990 introduction in Land Rovers, the Gemini diesel engine was then 'hijacked' as the principal engine for the new Discovery model, on which work had begun in 1986. There were some tense moments in the final development stages as the engineers struggled to get the engine ready for 1989 production and the Discovery's launch, but they succeeded. Discovery sales took off with a rush and by the time the engine was launched in the Defender, it had already been widely acclaimed around the world. It also had a new name. Land Rover's marketing people called it the 200Tdi: 200 was an approximation of its peak torque of 195lb ft, and Tdi stood for 'turbocharged, direct-injection, intercooled.'

In fact, the Defender version of the engine was not quite the same as that in the Discovery. The Discovery engine had its turbocharger mounted low down on the exhaust side; the Defender engine had its turbocharger mounted higher up because the low-mounted turbocharger fouled fixed items in the Land Rover engine bay. The Defender engine was also built in a slightly lower state of tune than the Discovery version. With a headline power figure of 107bhp as against the Discovery's 111bhp, it reinforced the idea that the Discovery was a superior kind of vehicle – and that suited Land Rover's marketing strategy perfectly.

The 200Tdi engine was such an instant success in the UK and in Europe that observers assumed it had replaced every other engine in the old Land Rover range. In fact, it did not; the 200Tdi was actually a direct replacement only for the Diesel Turbo, or Falcon engine. The old indirect-injection 2.5-litre naturally aspirated diesel remained available, as did the 2.5-litre 4-cylinder petrol engine and the 3.5-litre V8 petrol engine. However, the 200Tdi completely dominated sales in most territories and these other engines were generally specified only in Defenders for distant export markets. Behind all of these engines came the 5-speed LT77 gearbox with LT230R two-speed transfer gearbox. However, there were different internal ratios in both gearbox and transfer box to suit the power and torque characteristics of the different engines.

A comparison of the manufacturer's own figures for the 200Tdi and superseded Diesel Turbo engines helps to make clear just how much of an advance the new engine represented. In a Ninety, the Diesel Turbo had taken 22.5sec to reach 60mph (96km/h) from rest; in a Defender 90, the time was reduced to 15.7sec. The maximum speeds were 75mph (120km/h) and 84mph (135km/h) respectively and the fuel consumption was 23.7mpg (11.9ltr/100km) for the Diesel Turbo and 28.3mpg (10ltr/100km) for the 200Tdi.

Figures apart, the 200Tdi's greatest achievement was to make the Defender a much more viable everyday vehicle than the older Land Rovers had been. This gave it an appeal to a much wider group of buyers, most notably to those who had bought into the 4×4 boom of the 1980s and wanted to use such a vehicle instead of a conventional saloon or estate car. In the 1990s, Land Rover went on to build on this new appeal. Most obviously, it developed the Defender as a leisure and lifestyle vehicle as well as a workhorse model. Very noticeable was that the early 1990s – the period of the 200Tdi engine – saw Land Rover introduce its first special-edition models. In earlier times, the very idea of a special-edition Land Rover would have been laughed out of court.

This willingness to develop the Defender for leisure markets as well as for commercial users was also fundamental to the launch of special variants for the US market. There had been Land Rovers in the USA and Canada before of course, but sales had never been strong. By the early 1970s, they did not justify the cost of regular re-engineering to keep the vehicles compliant with constantly tightening US safety and exhaust emissions regulations, so in 1974 Land Rover had thrown in the towel and withdrawn from the North American markets. It had returned in 1987 with the Range Rover (by then already seventeen years old), and built up a good reputation through its subsidiary Range Rover of North America. Plans to follow the Range Rover's transatlantic success with other models soon followed and a special version of the Defender was introduced to North America in 1992 as a 1993 model. Thereafter, there were special North American Defenders until 1998, when legislation once again forced the vehicle out of the US market.

The 200Tdi engine also went on to earn an excellent reputation for reliability and toughness. Though superseded after just four years by a further development of the Gemini 4-cylinder, it is still highly regarded within the Land Rover fraternity more than two decades after its introduction.

Land Rover continued to sponsor the Camel Trophy adventure expedition up to 1998, but Defenders never featured as the crew vehicles. The event was used very much as a promotional tool, so the crew vehicles in this period were always Discoveries. There were nevertheless Defenders on every event in the role of support vehicles and these were painted in the Camel Trophy colours.

Right at the end of this period, but far too late to affect the Defenders of the 200Tdi era, the ownership of Land Rover changed again, as the Rover Group passed into BMW ownership.

The 1991 Model Year

Production of the 1991 models, the first Land Rover Defenders, began in July 1990. They were announced to the public at the NEC Motor Show in Birmingham on 18 September, then made their European début at the Paris Motor Show which opened on 2 October. The old size designators had been retained with the new name, so that there were Defender 90s and Defender 110s, but the ultra-long wheelbase model was now called a Defender 130. It still had the same 127in (3,226mm) wheelbase; Land Rover just thought the name sounded better.

The overall range was the same as it had been for 1989, with one exception. There was no County Station Wagon on the 90 chassis, which had been withdrawn mainly to protect sales of the Discovery. It was still possible to order a high-specification Defender 90 Station Wagon by plundering the options list, but some of the County features were simply not available. There were no carpets or leather-bound steering wheel, the four individual seats in the rear were not to be had, and nor were the body-coloured grille and wheel-arch eyebrows associated with the County specification.

Land Rover made sure that the new Defenders were readily recognizable by giving them new badges and distinctive side decals. The Defender name replaced earlier styles above the grille, while new tape badges reading Defender 90, Defender 110 or Defender 130 were used at the rear. Above them, the old plastic Land Rover oval badge was replaced by the three-dimensional green and gold Land Rover logo, as first seen on the Discovery. Side stripes carried the Defender name in large letters, or the Defender name with the County name where appropriate.

On those models with Tdi and V8 engines, there was a decal badge on each front wing, low down behind the wheel arch;

The original Defender side stripes were quite garish, as is apparent from this 1991-model Station Wagon. The Rostyle wheels were still an option on short-wheelbase models. The Tdi engine derivative badge can be seen behind the front wheel.

The graduated side stripes on County models were altogether more attractive and the door sections were angled the opposite way. This was a 110 County Station Wagon.

the 2.5-litre petrol and 2.5-litre naturally aspirated diesel engines brought no such badges. Two new colours (Grey and a different White) were also added to the options list and another new option was a tail door without a spare wheel mounting. Suitable only for hardtops and Station Wagons, it tidied up one area of the vehicle that had always looked wrong when the spare wheel was carried inside the body or on the bonnet. For most markets, including Europe, side repeater indicators on the front wings became a standard item.

There were valuable changes inside, as well. Although base models still had grey vinyl upholstery, the County variants had a new cloth upholstery, known as Moorland. This had a dog-tooth check pattern and was essentially grey, although it incorporated multicoloured yarns which helped it integrate better with the range of exterior colours. The 110 County models actually lost their rubber mats – supposedly to improve the interior appearance – but the real change was to the location of the front seats.

'Land Rover elbow' had been a problem since the marque's earliest days. It was a condition brought about by the driver constantly banging his elbow on the hard metal of the door and it was caused because the seats were uncomfortably close to the outside edge of the vehicle. With the Defender, Land Rover's designers finally grasped the nettle. They moved the outer pair of front seats inboard by an inch each. In so doing, they of course reduced the width available for a centre front seat, but it seemed that fewer and fewer customers were actually using that seat by this stage, anyway. So in its place, they standardized a cubby box – a new lockable type that incorporated a trinket tray and was in black to match the door trims. In cases where the bulkhead and tools would have been visible behind the cubby, a vinyl modesty curtain was fitted. It was still possible to order a centre front seat in place of the cubby box (though it was of course narrower than before) and a centre front seat remained standard on twelve-seater Station Wagons for the UK, where the number of seats still kept the vehicle free of additional tax.

Minor changes elsewhere made their contribution. All models except for soft-tops now had door-operated courtesy lights as standard (though they were rather flimsy items), with a three-position switch on the dashboard. There were also two new functions for the windscreen wipers, one being an intermittent setting and the other a flick-wipe.

Not surprisingly, there were no major changes during the 1991 model year. A raised engine air intake became an option in January 1991 and early in 1991 there was some rationalization of braking components between short- and long-wheelbase models. But that was all.

STANDARD COLOURS AND TRIMS — 1991 Model Year

Paints and Side Stripes

		Standard Side Decals	*County Side Decals*
Beige	LRC433	Two-tone green with red	Strobe Green
Blue	LRC424	Two-tone green with red	Strobe Blue
Green	LRC419	Two-tone green with red	Strobe Green
Grey	LRC476	Two-tone green with red	Strobe Blue
Red	LRC390	Two-tone green with red	Strobe Blue
White	LRC456	Two-tone green with red	Strobe Green

Side Stripes

Side stripes on non-County models had stripes of two-tone green and red with the words 'Defender 90' or 'Defender 110' as appropriate on the front doors. The County side stripes read 'Defender' and 'County' on the doors.

Upholstery and Trim

Grey vinyl upholstery was used on base models, while County derivatives had Moorland grey dog-tooth check cloth upholstery.

The 1992 Model Year

The 1992 model-year Defenders had very few changes from the 1991 models and were available with the same range of exterior colours. They were introduced late by Land Rover standards, being announced at the Royal Smithfield Agricultural Show in December 1991.

There were some changes, though, the most obvious of which was a set of new side stripes for the non-County models, mercifully less garish than the originals. All were in two-tone grey, with the Defender name in yellow and a smaller County name at the rear when appropriate. The 1992-model Defenders for the UK came with power-assisted steering as standard rather than an extra-cost option and this was accompanied by a smaller steering wheel with a new design. There were also some changes to the optional sunroof. The 2.5-litre petrol engine disappeared from the UK price lists, although it would remain available for several more years as an export option. The synchromesh in the LT77 gearbox was improved, turning it into an LT77S type, and as so few Defender users ever troubled with a starting handle, this was deleted from the tool kit and a new front bumper without a cut-out was standardized.

There was a new design of side stripe for the 1992 models, as seen here on a 90 hardtop with the optional Rostyle wheels.

Land Rover continued to capitalize on the Tdi engine's strengths, too. During 1992, they introduced a conversion kit for older coil-spring models, making it available through Land Rover Parts. The main benefits were to fuel economy, maximum speed and towing ability, but it was the combination of all three that would have been decisive. Fuel consumption improvements alone were not great and the predicted economy for a converted One Ten Station Wagon was 19.8mpg (14.3ltr/100km) as opposed to the 17.8mpg (15.9ltr/100km) it would have recorded with its original 2.5-litre petrol engine.

More far-reaching than that, however, was the arrival in March 1992 of the first special-edition Defender. Special editions were a marketing tool and were quite common in the passenger car market. They could have many purposes: they could test customer reaction to a new feature; they could provide a rapid counter to a rival maker's new model by offering more features at a similar price; or they could improve sales through an association with something that had become popular or topical. Land Rover had used them before, but only on the Range Rover – there had been In Vogue special editions in 1981, 1982 and 1983, and in Germany there had been a special Camel Trophy edition of the Range Rover in 1981. Never before, though, had there been a special edition of the workhorse Land Rover.

That first special edition in March 1992, perhaps surprisingly, was not for the UK market. Instead, the Land Rover Trophy was released in France, and by all appearances it had been created by Land Rover France, who had simply added extra features from the options list to a standard Defender 90 Station Wagon. The addition of 'exclusive' features not available on other models helped to give the Defender 90 Trophy its special appeal, although it must be said that the special features were limited to a pair of decals on the doors and some side stripes, so were hardly enticing in themselves!

Steel disc wheels were still standard on commercial derivatives of the Defender 90, as this 1992 truck cab shows.

Land Rover press pictures were a regular source of confusion in the late 1980s and early 1990s. This 1992 110 Station Wagon was described as a County type – which it patently was not.

STANDARD COLOURS AND TRIMS

1992 Model Year

Paints

Beige	LRC433
Grey	LRC476
Blue	LRC424
Red	LRC390
Green	LRC419
White	LRC456

Side Stripes

Side stripes were in two-tone grey with 'Defender' in yellow-beige on each front door and 'County' as appropriate on the rear side panels.

Upholstery and Trim

Grey vinyl upholstery was used on base models, while County derivatives had Moorland grey dog-tooth check cloth upholstery.

Nevertheless, Land Rover France had clearly understood its market very well and would be responsible for many more special-edition Defenders over the years that followed. Chapter 10 looks more closely at the multiple special editions that have been released over the years.

The 1993 Model Year

The 1993 model year would see some important developments in the Defender range, including the arrival of a special North American model, the expansion of the County range and a special-edition model for the leisure market. One of those special-edition models (possibly the first, or a prototype) would become the ceremonial vehicle for the July 1993 event to celebrate what Land Rover called the 1.5 Millionth Defender. It was in fact no such thing: the 1.5 million was the number of workhorse Land Rovers built since 1948 – and the figure was very much an approximation, thanks to inaccuracies in the records.

The 1993 model year began in autumn 1992, but its features reached production in several stages. From the start, Beige was deleted from the colour options and in September a 65-amp alternator replaced the earlier 45-amp type, giving greater reserves of electrical power and reflecting the increasing number of electrical items typically being fitted on the latest Defenders. Hardtops and Station Wagons were readily recognizable as 1993 models because they now had a white plate covering the centre of the spare wheel mounted on the tail door. This met new European regulations about dangerous protrusions. Windscreen hinges were also deleted at about this time, as windscreens generally were no longer required to fold downwards onto the bonnet.

Then between September and November, a Noise Reduction Package for the 200Tdi engine was introduced in stages: good though it was, the 200Tdi was still a fairly vocal engine. First came lower-rate engine mounts (on production from 7 September), then extra insulation for the dash in the toe box area (from 12 October), and finally a stiffer propshaft, for 110 models only from late November.

November then brought the changes for the County models, which could be had with metallic paints or a clear coat over base black for the first time. The new colours, available from 15 November, were shared with the Range Rover and Discovery ranges, where they had suitably exotic names, but it was still policy to call a spade a spade for the Defender range. So they became County Black, County Green and County Red. County models for the 1993 model year still had body-colour wheel-arch eyebrows, but the 110s reverted to black grilles and headlamp surrounds. Wheels were white and so was the roof on standard examples, but for the first time it became possible to order the roof in the same colour as the body. There was an extra charge, of course.

For the mainstream Defenders, a whole set of additional changes followed in spring 1993, being introduced over a period of around three months. Collectively, Land Rover knew them internally as the 93.5 model year changes. Most visibly, the side stripes were deleted from all non-County models;

The 1.5 millionth 'Defender' was, of course, the 1.5 millionth utility Land Rover since 1948; Land Rover had not managed to build 1.5 million Defenders in less than three years! The ceremony was held at the Solihull plant in July 1993 and the VIP guest who actually drove the vehicle off the assembly line was Canadian rock singer and Land Rover enthusiast Bryan Adams, seen here in white on the left.

New on the 1993 models was a cover for the spare wheel recess. It met legislation about potentially harmful projections. This 110 Station Wagon has no side stripes; customers for commercial derivatives were generally not keen on them and often asked for them to be left off.

Side stripes were standard on commercial derivatives until the middle of the 1993 model-year. This 90 hardtop has the optional Rostyle wheels as well.

it appears that customers had actually not liked them. Most importantly, and beginning with Defenders built on 4 May, the rear drum brakes were replaced by discs, as seen on the NAS 110 (see below), and asbestos-free pads were standardized all round. The effect, realistically, was minimal, but Land Rover needed the all-disc specification to make the Defender look as modern as its obvious rivals.

The County range was expanded, too. To the existing 110 Station Wagon, still available with either ten or twelve seats, were added two models. The 90 County Station Wagon returned to the lists after a three-year absence, as the need to protect Discovery sales had now passed. It could be had with either the Tdi diesel or V8 petrol engine, although V8s remained rare. The second new model represented a move into a new area of the market; it was a 90 County Hardtop. Visually, it was distinguished by its body-colour wheelarch eyebrows, but it came with Moorland cloth seats and other County appointments. It also had its own extra-cost

County models still came with side stripes as standard and were distinguished from lesser models with similar side stripes by the word 'County' on the rear flanks.

STANDARD COLOURS AND TRIMS

1993 Model Year

Paints

Blue	LRC424	
County Black	LRC416	Same as Range Rover Beluga Black and Discovery Caracal; special order only
County Green	LRC413	Same as Range Rover Ardennes Green and Discovery Carrigada Green
County Red	LRC467	Same as Range Rover Trocadero Red and Discovery Foxfire Red
Green	LRC419	To 5 July 1993 only
Grey	LRC476	
Red	LRC390	
White	LRC456	

Side Stripes

Side stripes were in two-tone grey with 'Defender' in yellow-beige on each front door and 'County' as appropriate on the rear side panels. Stripes were deleted from non-County models in April 1993.

Upholstery and Trim

Grey vinyl upholstery was used on base models, while County derivatives had Moorland grey dog-tooth check cloth upholstery.

Accessory Pack, consisting of a wrap-around nudge bar, side runners and rear lamp guards.

There were then some changes to the colour options during June and July, as the green was replaced by a new shade and an additional metallic red was added to the County options. Yet none of this prepared the public for the special edition that was to be launched in July. The SV90, as it was called, made clear that Land Rover had every intention of pushing the Defender further and further into the leisure-vehicle market in the hope of additional sales. It was a summer-weekend special, an open model with padded rollover cage, special paint scheme and chunky off-road tyres on the first alloy wheels ever seen on a Land Rover. Most were actually sold after the 1994 model year had begun in the autumn, but a few were genuine 1993 models. Chapter 10 has more about it.

At the start of the 1993 model year, Land Rover also announced a special-edition Defender 110 for North America. There had been no Land Rovers for North America since 1974 (although Range Rover had gone on sale there in 1987), and this model was the start of a determined assault on the North American market. When it went on sale on 1 August 1992 through the company that had been Range Rover of North America, that company changed its name to Land Rover North America. The next stage – although this had not been publicly revealed at the time – would be to broaden the North American product range by importing the Discovery to the USA and Canada.

The NAS 110

The task of the NAS 110 (North American Specification 110) was to make the image of the Defender as a tough, no-nonsense 4×4 as widely appreciated in North America as it was elsewhere. It was a difficult task and a great deal of work had gone into preparing the vehicle to do it.

The first prototype was a modified standard production LHD Station Wagon, registered as J31 SKV. There were then nine more prototypes (numbered as LRNA/1 to LRNA/9) that were used for extensive testing. One of these, which differed in several details from the eventual production models, was shown to the media at a special event on 8 January 1992, just before the Detroit Motor Show opened. It was not, however, put on display at the Show itself.

The production models were partially built on the main assembly lines and were then sent to Land Rover Special Vehicles (LRSV), newly reorganized from its time as Special Vehicle Operations, where they were hand-finished with their special features. They were built in batches and their assembly was seen as a proving operation for LRSV, which would later take on responsibility for small-run limited-edition models of all kinds and for the Autobiography custom-finishing scheme, as well as for the specialist commercial conversions which Special Vehicle Operations had built.

The NAS 110 was powered by the same injected 3.9-litre V8 petrol engine as was then used in contemporary Range Rovers for the USA; this had already been emissions-certified and had catalytic converters in the exhaust. It had 180bhp at 4,750rpm and 227lb ft at 3,500rpm. The gearbox was an LT77S 5-speed manual, with the usual LT230 two-speed transfer box, and there were disc brakes on the rear wheels as well as the fronts – an exclusivity of which its North American customers, not being used to existing Defenders, were blissfully unaware.

The vehicles were all built as nine-seater Station Wagons, with a special cubby box between the two front seats and four individual inward-facing seats in the rear. Upholstery was in County-style Moorland cloth and the passenger compartment was fully carpeted. The special cubby box contained a US-specific Pioneer AM/FM stereo radio-cassette head unit and there were four speakers in the cabin. All examples had in-dash air-conditioning and a special dash panel that contained a clock, an ammeter, an oil-pressure gauge, a cigarette lighter and the switch for the rear window wash-wipe. A heated windscreen was also part of the standard specification, together with the auto-dimming rear-view mirror used in contemporary Range Rovers.

As the Defender's body was not strong enough to meet US rollover regulations, the NAS 110s had a tough external roll cage, made in the UK by the specialists, Safety Devices. This was painted white to match the body, which was white with a white roof and wheels. Contrasting black was used for the wheel-arch eyebrows, the standard side runners and the wrap-around bull bar mounted on a special front bumper.

This bumper had amber indicator lights inset in its front face. There were also amber front running lights, amber side markers to meet US regulations and special rectangular tail lights in place of the standard round items, because these were too small to meet US regulations. A third brake light was mounted behind the glass of the tail door's heated rear window, again to meet US regulations. There was also a bumper bar on the rear cross-member, angled to follow the lower edge of the cross-member.

To give the vehicles exclusivity and so additional buyer appeal, the NAS 110 was marketed as a numbered limited edition. Each vehicle had an oval plate on the rear, inscribed with its number within the edition: US models ran from 1/500 to 500/500, while Canadian models were numbered CDN1/25 to CDN25/25.

As Land Rover had hoped, the NAS 110s were snapped up by wealthy buyers who wanted something different. Among them were a number of high-profile celebrities who gained precisely the publicity for the vehicle that Land Rover wanted. One insisted on having his vehicle finished in

Riding on the success of the Range Rover, Land Rover developed a special Defender 110 derivative for the North American market.

LEFT: **This was one reason why commercial buyers were not keen on the side stripes: they tended to detract from their own signwriting on the vehicles.**

BELOW: **This 1994 Defender 130 has a Quadtec 4 box body built by LRSV.**

all-over black, but the rest had the standard white paint. However, the NAS 110 proved to be much more than an overnight fashion sensation. More than twenty years after they were built, examples still change hands for high prices. Even the original sales brochure is a prized item.

The 1994 Model Year

The 1994 model year was a short one for the Defender. It lasted only for the six months from September 1993 to March 1994 and was treated more or less as an extension of the 1993 model year. Other matters were occupying Solihull's resources this year – the Discovery was the main focus of attention in autumn 1993 as it took on the 3.9-litre V8 engine and an automatic gearbox option for diesel derivatives. At the same time, the Autobiography custom-finishing scheme for Range Rover was announced. Nevertheless, the Defender range took an important step forwards with the introduction of the 90 for North America.

Things were quiet in the UK, though. The late arrival of the 90SV meant that marketing focus was on this at the time of the London Motor Show in September 1993, and there was only one other Defender on the Land Rover stand. This was a 110 County ten-seat Station Wagon, not visibly changed from the previous model year's specification. Just one important new option was introduced, and that had

been listed since June 1993. Bearing the rather cumbersome name of Freestyle Choice, it brought five-spoke alloy wheels for 90 County models.

In January 1994, Land Rover Parts announced the availability of an addition to its Tdi retro-fit kits for Defenders. This time, a kit was made available to suit V8-engined 110s with the LT77 gearbox. In practice, that meant Defender 110 V8s built in the 1991 and subsequent model years. Advertising suggested that fuel economy would improve by a staggering 58.3 per cent after the conversion was done.

The NAS 90 (*left*) was an eye-catching machine and could be had in eye-catching colours, too. This was AA Yellow. Apparent in this picture is the rear-mounted fuel filler; tank and filler would not move to the rear on standard-production Defender 90s until the Td5 era. The upholstery (*below*) was a special Twill PVC, and the presence of air-conditioning controls in the centre of the dash meant that the radio had to move to the cubby box between the seats.

The NAS 90

The 1994-model Defender 90 for North America, or NAS 90, was announced on 3 September 1993 for October availability. Its whole concept was deliberately sporting and was designed to tackle the 'weekend fun' market in the USA that Jeep had developed so effectively since the 1960s. The NAS 90 had a 3.9-litre V8 engine and LT77S 5-speed manual gearbox, driving through a two-speed transfer box. As on contemporary Range Rovers, the engine had redesigned pistons and was slightly more powerful than the V8 in the NAS 110, with figures of 182bhp and 232lb ft.

Despite similarities to the UK-market SV90, the NAS 90 was different in many respects. It had military-style doors with removable tops and recessed handles, while the 90SV had standard full-height doors. The North American vehicle had Freestyle Choice five-spoke alloy wheels rather than the flat-spoke type, and they were shod with the chunkier tyres that the SV90 team had wanted for their vehicle as well – BF Goodrich 265/75R16 Mud Terrains. These were theoretically optional, but all vehicles had them. Front and rear anti-roll bars were part of the package. The fuel tank was also now at the rear, with a filler behind the rear wheel.

A range of five bright colours was available, all with the evocative names expected in the USA, and the NAS 90 came with the same front bumper and bull bar as the 1993-model NAS 110, plus the same rectangular rear light units. Instead of a Defender badge above the grille, there were silver plastic 3D letters, reading 'Land Rover'. There was the same swing-away spare wheel carrier as on the 90SV and each NAS 90 had a numbered plaque on the tail panel.

The front 'safari' roll cage was standard, but a full cage covering the rear was also available. There was a flimsy 'Bimini' canvas half-top to provide some protection for the passengers in front, and two additional seats for the rear were an extra-cost option. The Tickford-designed soft-top was another option for the rear, but many gave trouble in service and were replaced by a soft-top designed in the USA by Bestop.

There were normally two seats, upholstered in Charcoal 'twill-effect' vinyl with a cubby box between them. This had a lockable radio security cover and the radio was a US-specific Clarion stereo unit, with AM/FM radio and a cassette player, plus four speakers that were claimed to be rain-resistant. Rubber floor mats were fitted. Air-conditioning was an extra-cost option, with LRNA also offering a range of accessories for the vehicle.

Sixty examples (all in Coniston Green) were sold with a locally made removable hardtop. Towards the end of the season, a run of 100 specially-equipped models in Beluga Black was put on sale. These had an A-frame protection bar and side runners in contrasting stainless steel. They also came with leather upholstery and a Surrey top.

There were 1,943 of the 1994-model NAS 90s for the USA, plus a further eighty-seven for Canada, making 2,030 in all. The NAS 90 was voted 'Four Wheeler of the Year' by the US off-roading magazine *Four Wheeler*.

The Defender 130 was still being built as a Crew Cab model, although quite a high proportion were ordered with special bodywork by LRSV.

STANDARD COLOURS AND TRIMS

1994 Model Year

Paints

Blue	LRC424	
Coniston Green	LRC637	Replaced Green from 5 July 1993
Grey	LRC476	
Red	LRC390	
White	LRC456	
County Black	LRC416	Same as Range Rover Beluga Black and Discovery Caracal; special order only
County Green	LRC413	Same as Range Rover Ardennes Green and Discovery Carrigada Green
County Red	LRC467	Same as Range Rover Trocadero Red and Discovery Foxfire Red
Montpellier Red	LRC536	Added to County options from 21 June 1993

Paint Colours for the 1994 NAS 90

AA Yellow	LRC559	
Alpine White	LRC456	
Arles Blue	LRC424	
Beluga Black	LRC416	100 examples only, in summer 1994
Coniston Green	LRC570	
Portofino Red	LRC390	

Side Stripes

County models had side stripes in two-tone grey with 'Defender' in yellow-beige on each front door.

Upholstery and Trim

Grey vinyl upholstery was used on base models, while County derivatives had Moorland grey dog-tooth check cloth upholstery.

VINS

(i) European-Type VINs

The VIN codes used in this period consisted of seventeen characters, made up of an eleven-character prefix code and a six-digit serial number.

Example: SALLDHAF7HA-xxxxxx

This breaks down as follows:

SAL	Manufacturer code (Rover Group)
LD	Defender
H	110in (2,794mm) wheelbase
	K = 127-inch wheelbase (Defender 130)
	R = 110in (2,794mm) wheelbase
	S = 110in (2,794mm) wheelbase (military specification)
	V = 92.9-inch wheelbase (Defender 90)
A	Truck cab, hardtop or soft-top body
	B = Two-door Station Wagon (Defender 90)
	E = Two-door Crew Cab (Defender 110 and 130)
	F = Four-door Crew Cab (Defender 110 and 130)
	H = High-Capacity Pick-Up (Defender 110)
	M = Four-door Station Wagon (Defender 110)
F	2.5-litre 200 Tdi diesel engine
	C = 2.5-litre diesel engine
	D = 2.5-litre petrol engine
	L = 3.5-litre V8 carburettor petrol engine
7	RHD with 5-speed manual gearbox
	8 = LHD with 5-speed manual gearbox
H	Model year 1991
	J = 1992
	K = 1993
	L = 1994
A	Assembled at Solihull
	F = Assembled overseas from CKD
xxxxxx	Serial number
	700001–800000 range for Defender 90
	800001–900000 range for CKD models
	900001–999999 range for Defender 110

(ii) NAS VINs

The VIN codes used in this period consisted of seventeen characters, made up of an eleven-character prefix code and a six-digit serial number.

Example: SALDH1281PA-xxxxxx

This breaks down as follows:

SAL	Manufacturer code (Land Rover)
D	Defender
H	110in (2,794mm) wheelbase
	V = 90in (2,286mm) wheelbase
1	Four-door body
	2 = Two-door body
2	3.9-litre V8 injected petrol engine
8	LHD with 5-speed manual gearbox
1	Security check digit
	(0 to 9, or X)
P	Model year 1993
	R = 1994
A	Assembled at Solihull
xxxxxx	Serial number
	915000-922999 range for Defender 110
	930000-948999 range for Defender 90

Engine Identification Numbers for Defender, 1991–4 models

Engines have a nine-digit identification number. The digits break down into three groups: the first group of three digits identifies the engine type, the second group of five digits is the engine serial number and the final digit identifies subgroups created by running specification changes. Example: 24G-00001A.

2.5-litre diesel engine

12J-00001A	Standard engines
18J-00001A	For British MoD vehicles, with 24V electrical system

2.5-litre petrol engine

17H-00001A	Standard engines
18H-00001A	Military engines with 24V electrical system

VINS AND ENGINE NUMBERS 1991–4 Models

19H-00001A With auto gearbox, for unsuccessful Swiss military tender

2.5-litre 200 Tdi diesel engine

11L-00001A Standard engines

3.5-litre V8 carburettor engine

20G- xxxxx C 8.13:1 compression, non-detoxed

24G- 00001A 8.13:1 compression, detoxed

3.9-litre V8 injection engine

37D-00001A 8.13:1 compression, with catalytic converter exhaust

SPECIFICATIONS DEFENDER 1991–4

Petrol Engines

Type	V8
Bore and stroke	88.9mm x 71.1mm
Capacity	3,528cc
Valves	V8-cylinder with pushrod ohv
Compression ratio	8.13:1
Fuel system	two Zenith-Stromberg carburettors
Max. power (DIN)	135bhp at 5,000rpm and 182lb ft at 2,500rpm

Type	NAS V8, 1993–4
Bore and stroke	93.98mm x 71.1mm
Capacity	3,947cc
Valves	V8-cylinder with pushrod ohv
Compression ratio	8.13:1
Fuel system	Lucas L-Jetronic fuel-injection
Max. power (DIN)	180bhp at 4,750rpm and 227lb ft at 3,500rpm (110 models); 182bhp at 4,750rpm and 232lb ft at 3,100rpm (90 models)

Diesel Engine

Type	200Tdi
Bore and stroke	90.47mm x 97mm
Capacity	2,495cc
Valves	4-cylinder with ohv
Compression ratio	19.5:1
Fuel system	direct-injection and Garrett T25 turbocharger
Max. power (DIN)	107bhp at 3800rpm and 188lb ft at 1,800rpm

Transmission

Permanent four-wheel drive

Main gearbox

(a) 1991–2 model years, with 2.5-litre petrol and 2.5-litre naturally aspirated diesel engines:

5-speed LT77 manual gearbox

Ratios 3.585:1, 2.301:1, 1.507:1, 1.000:1, 0.831:1, reverse 3.701:1

(b) 1991–4 model years, with V8 petrol and 200 Tdi turbodiesel engines:

5-speed LT77 manual gearbox (LT77S with improved synchromesh from 1992 model year)

Ratios 3.692:1, 2.132:1, 1.397:1, 1.000:1, 0.770:1, reverse 3.429:1

Transfer gearbox

(a) 1991–2 model years, with 2.5-litre petrol and 2.5-litre naturally aspirated diesel engines:

2-speed LT230

Ratios 1.667:1 (high) and 3.320:1 (low)

(b) 1991-1994 model years, with V8 petrol and 200 Tdi turbodiesel engines:

2-speed LT230

Ratios 1.411:1 (high) and 3.320:1 (low)

Continued overleaf

Continued from previous page

SPECIFICATIONS — DEFENDER 1991–4

Suspension and Steering

Suspension

Coil springs all round

Steering

Worm and roller steering with 20.2:1 ratio; optional power assistance with 19.3:1 ratio. PAS standard from January 1992.

Brakes

Type Disc brakes on the front wheels and drum brakes at the rear (1991–3 models); disc brakes on all four wheels (1993 on); some heavy-duty models with ventilated front discs; servo assistance; separate drum-type transmission parking brake

Electrical System

12V with alternator

Dimensions

90 models

Track	58.5in (1,486mm)
Wheelbase	92.9in (2,360mm)
Overall length	146.5in (3,720mm)
Overall width	70.5in (1,791mm)
Unladen height	77.3in (1,963mm) for truck cab 77.4in (1,966mm) for soft-top 77.6in (1,971mm) for hardtop and Station Wagon
Unladen weights	3,587lb (1,627kg) minimum with V8 petrol engine 3,607lb (1,636kg) minimum with 4-cylinder petrol engine 3,651lb (1,656kg) minimum with naturally aspirated diesel engine 3,734lb (1,694kg) minimum with Tdi diesel engine

110 Models

Track	58.5in (1,486mm)
Wheelbase	110in (2,794mm)
Overall length	175in (4,445mm) for truck cab and soft-top 180.3in (4,580mm) for Station Wagon 184in (4,674mm) for HCPU
Overall width	70.5in (1,791mm)
Unladen height	80.1in (2,034mm)
Unladen weights	3,979lb (1,805kg) minimum with 4-cylinder petrol engine 3,981lb (1,806kg) minimum with V8 petrol engine 4,030lb (1,828kg) minimum with naturally aspirated diesel engine 4,127lb (1,872kg) minimum with Tdi diesel engine

130 Models

Track	58.5in (1,486mm)
Wheelbase	127in (3,226mm)
Overall length	198in (5,029mm)
Overall width	70.5in (1,791mm)
Unladen height	80.1in (2,034mm)
Unladen weights	4,436lb (2,012kg) minimum with V8 petrol engine 4,599lb (2,086kg) minimum with Tdi engine

Performance

90 NAS soft-top with 182bhp V8 and 5-speed LT77

0–60mph	10.2sec
Maximum	86mph (138km/h)
Fuel consumption	14–15mpg (US gallons)

110 Station Wagon with 134bhp V8 and 5-speed LT85

0–60mph	15sec
Maximum	84mph (135km/h)
Fuel consumption	13–15 mpg (21.8–18.9ltr/100km)

110 with 200Tdi engine and 5-speed LT77

0–60mph	Not quoted
Maximum	Not quoted
Fuel consumption	29mpg (9.8ltr/100km)

CHAPTER SEVEN

THE CLASSIC ERA, 1995–1998

The 300Tdi period, 1995–8 Model Years

The 200Tdi engine was inclined to be noisy, despite the measures Land Rover had taken to quieten its combustion rattle, and the LT77S gearbox could be rather baulky when cold unless it was filled with automatic transmission fluid. So at the start of the 1990s, Land Rover embarked on engineering programmes to improve the Tdi engine and to design a new gearbox. It was this new power train that defined the next period of the Defender story.

For a complex set of reasons, partly associated with the introduction of airbags on the Range Rover and the Discovery, and with the introduction of the Discovery to the US market, the 1995 model year started in March 1994. That was when the new Defenders were introduced, too. The newly improved diesel engine was branded 300Tdi to distinguish it from the earlier 200Tdi. The new gearbox was called the R380 – R for Rover, and 380 for its maximum torque capacity in Newton-metres (around 280lb ft in Imperial measures).

Work had started on refining the Gemini engine almost as soon as it had been signed off for production in the first few months of 1989. The more refined engine was known within Land Rover as Gemini 2. In the meantime, a separate group of engineers was looking at militarizing the engine, as it was obvious that the new diesel would be needed to meet military orders. The militarized Tdi programme was known as Project Vixen and its key feature was a redesigned front end that would allow room for the big 90-amp generator typically used on military models fitted with the 24V electrical system needed to operate two-way radios.

In due course, the two programmes were amalgamated and the new engine to suit both sets of requirements was baptized Gemini 3. It was this engine which was introduced in March 1994 for the Defender – and it went into Range Rovers and Discoveries at the same time, too. It was the 3 of Gemini 3 which inspired its new type designation of 300Tdi; sadly, the new figure did not imply a massive increase in the torque output!

In fact, the 300Tdi had been deliberately designed to achieve the same power and torque outputs as the Range Rover and Discovery versions of the outgoing 200Tdi. This meant 111bhp at 4,000rpm and 195lb ft at 1,800rpm for all models. This time, there was no special derivative for the Defender and the workhorse model used the same engine as the other models, with the same high-mounted turbocharger. Working with more time than had been available when the 200Tdi had been used to create the original Defender, Solihull's engineers had moved the engine slightly forwards in the chassis to allow the turbocharger to clear the Defender's left-hand front suspension tower.

Most of the elements of the original Tdi engine remained in place – it retained the same bore and stroke, and its single camshaft was still driven by a toothed belt. However, no fewer than 208 components had been changed. Among others, the 300Tdi had a new cylinder head, fuel injectors, pistons, conrods, turbocharger, exhaust manifold, timing belt, water pump and alternator. Perhaps of most importance to Land Rover was that the redesigned engine also took less time to build than its 200Tdi predecessor, so reducing manufacturing costs. The need to meet new European emissions regulations for diesel engines scheduled to come into force for 1996 had also been taken into account during the development process and the engines were capable of complying with the new rules from their introduction in 1994.

As far as the customers were concerned, the changes that made the 300Tdi brought about a huge increase in refinement. Although the engine was still a long way from being as quiet as

The 300Tdi was a very much refined evolution of the 200Tdi type. Defender versions had the same high-mounted turbocharger as Discovery and Range Rover derivatives.

the best indirect-injection types from other manufacturers, it was very much less vocal than the 200Tdi type, while still offering the same outstanding fuel economy that rival indirect-injection types could not match. The 300Tdi would go on to become one of Land Rover's best-loved engines. Apart from some problems with timing belts, which were cured by a modification after the engine had gone out of production in the UK, it proved to be a durable and reliable power unit. Most important, it did not depend on complex electronic controls and could therefore be repaired simply without the aid of special tools after a breakdown in remote areas.

As for the R380 gearbox, Land Rover's aim had been to make a manual box with a change as slick and easy as the best of its Japanese rivals. In that, the engineers were certainly successful and the reduced clutch pedal pressures that accompanied the new gearbox in its Defender installation were also welcome. Nevertheless, the R380 gearbox suffered a number of failures in its early days, when poor quality control on the assembly lines led to many examples jamming in gear.

Once that had been sorted out, the R380 became well liked. It had a different gate pattern from the LT77S it replaced, a double-H with reverse behind fifth replacing the older type with reverse on a spur alongside first. It took seasoned Land Rover owners a long time before they could trust themselves to select fourth rather than reverse when changing down from fifth, even though there was a strong détente spring to prevent such problems.

This was an era in which Defender development was driven as much by new legislation as by customer demands or engineering advances. Most notably, there were the 'Euro 2' European Community diesel emissions regulations from 1 January 1996 and a very complex set of requirements governing common safety standards from 1 January 1998.

The 1995 Model Year

The new 300Tdi/R380 driveline was the major change for the 1995 model year which, as noted above, began in March 1994. In the UK, Tdi diesel derivatives continued to account for almost all sales, although V8 petrol County Station Wagons were still listed. These, too, had the new R380 gearbox. The 2.5-litre petrol engine and 2.5-litre naturally aspirated diesel were still available for export territories beyond Europe.

There were only minor changes in other areas of the vehicle. Colour and trim options remained unchanged. Land Rover claimed to have worked on improving the reliability of the electrical system and there were several new interior features that were intended to improve user convenience. The actual changes brought improved interior lighting, with the courtesy light switch now on the light itself; there was a new clock with a less cluttered appearance; from Rover Cars came the tried and tested components for an improved handbrake warning light system; and there was a new ICE system with upgraded radio-cassette unit and speakers. Station Wagons for 1995 also came with belts for the inward-facing rear seats as standard.

Land Rover's 1995 model year began much later in the USA, because it was timed to coincide with the availability of the new 38A-model Range Rover for that country. So although press releases about the 1995 Defender 90 went out in late December 1994, the vehicles did not go on sale until January – a full ten months later than the start of the 1995 model year in Europe. They were essentially the same as the 1994 NAS 90s, but incorporated the new R380 gearbox, US-made soft-tops (LRNA had lost patience with the Tickford type), and a few other refinements prompted by customer feedback. The open model was also accompanied by a Station Wagon derivative, which was primarily intended to aid sales in the colder northern states of the USA. However, slow sales of the 1994 models dissuaded the Canadian dealers from taking any of the 1995 models.

The 1995 model year was also when a very special version of the Defender 90 was released for the specific use of the Carabinieri in Italy. The Carabinieri is a military force with functions similar to those of the French gendarmerie, but with a dual function as civilian and military police. By the early 1990s, it needed a new light 4×4 for general duties, but the Fiat Campgnola, on which it had relied for many years was now out of production, and Land Rover Italy was able to make a strong pitch and offer a version of the Defender 90 as a replacement.

It was to be a large contract, for 700 vehicles, but there was no 90 derivative at the time that fitted the bill. The Carabinieri wanted a petrol-powered Station Wagon, but the V8 engine was not considered suitable, probably mainly because its capacity put it into the luxury class tax bracket in Italy. Keen to win the contract, Solihull agreed to provide a special model with the 2.0-litre T16 petrol engine. This was a 4-cylinder, 16-valve engine that had originally been developed for transverse installation in Rover cars, but Land Rover had already engineered a north–south installation for the Discovery Mpi in 1993.

The T16 engine had 134bhp, which made it only slightly less powerful than the final versions of the 3.5-litre

ABOVE: **Rostyle wheels disappeared from the options for the 90. Instead, there were five-spoke alloys, which were accompanied by anti-roll bars in a package rather awkwardly described as Freestyle Choice. This is a 90 County Station Wagon, on which they were standard. The winch, of course, was not.**

BELOW: **Despite the ever-increasing emphasis on the passenger-carrying models, there was still no shortage of orders for commercial Defenders. This is a 110 High-Capacity Pick-up from 1995.**

carburettor V8, but its power and torque delivery were very different. It was a high-revving engine, so Solihull developed a special transmission for it, combining the Discovery Mpi version of the 5-speed LT77S gearbox with the transfer box of a Defender Tdi. All this was installed in a chassis that was essentially the same as that of the NAS 90, with that model's plastic fuel tank, fuel pipes, evaporative loss system and sensors. To make room for the Mpi engine's exhaust catalyst, however, a special cross-member was necessary. The vehicles were not assembled by Special Vehicles, but were built on the main assembly lines at Solihull, although final fitting out was done in the Carabinieri's own workshops in Italy.

Deliveries began in March 1995 at a rate of forty-five vehicles per month; the Carabinieri liked the vehicles enough to increase its original order from 700 to 840 vehicles. This made the order for these 2.0-litre 90s the largest-ever single order for Defenders from Italy and put its value up to £12 million. It also cemented the Defender 90 as a Carabinieri favourite, with the force continuing to buy Defenders after the completion of the original order, even though a policy change enabled them to switch to 300Tdi diesel models.

The 2.0-litre petrol engine delivered quite lively acceleration as well as reasonable fuel economy, and there was some discussion at Solihull about the possibility of using it in a sporty model otherwise similar to 1992's SV90. However, this idea was abandoned when sales of the Discovery Mpi proved to be poor.

The Italian Carabinieri bought a fleet of Defender 90s with the T16 petrol engine as a special fit. They later bought conventional Tdi-powered examples as well.

The 1995 NAS 90

There were two derivatives of the Defender 90 for US customers in the 1995 model year. One was an updated version of the open model sold during the 1994 season; the other was a six-seat Station Wagon. Both models had the 182bhp 3.9-litre V8 petrol engine, now driving through the latest R380 5-speed manual gearbox and LT230T transfer gearbox. A new feature on all the 1995 models was an engine immobilizer.

On all models, there were large round tail lights instead of the rectangular ones used on 1993 and 1994 NAS models, plus a matching amber turn-signal lamp at the front. The Mud Terrain tyres on 1994 models had given way to less aggressive BF Goodrich All Terrains, but the Freestyle Choice five-spoke wheels remained as a theoretical option; in fact, all vehicles seem to have had them. There was a more powerful ICE system and from mid-season an extra-cost CD changer and amplified subwoofer became available as a package. Extra stowage was added, in the form of pockets on the seat backs and doors and the sound insulation was improved. There were also two brush-bar options.

The open models again had Charcoal twill-effect vinyl upholstery. They came with a removable fastback-style 'Bimini' soft-top as standard, now made by Bestop, a US specialist. Bestop also made the full soft-top, an extra-cost option that used a rail fastening system instead of the troublesome press-studs of the 1994 Tickford top and was sometimes called a Surrey top because its sides could be rolled up. A further alternative was a removable Glass-Reinforced Plastic (GRP) hardtop, like that seen on some 1994 models and always finished in Ash Grey. For easy storage, the side and rear sections could be stowed within the roof section when the unit was removed.

The Station Wagons, fewer in number, had an external roll cage at the front and a full internal cage in the rear. A sunroof and heated rear window with wash-wipe system were standard. They were equipped as six-seaters, with a cubby box between the two front seats. Upholstery was described as Ash Grey houndstooth cloth, but was actually the standard County option of Moorland cloth. Full carpeting came as standard.

There were 1700 NAS 90s for 1995, rather fewer than in the 1994 season because of the shorter model year. Of these, 1,190 were soft-tops and 510 were Station Wagons. The range of colour options was also smaller than in 1994 – five instead of six – as LRNA adapted to the demands of a very different group of customers from those to whom it was selling Range Rovers and Discoveries.

The NAS 90 was much the same for 1995 as it had been for 1994. Among the options was a removable GRP hardtop, and when the full roll-cage was fitted, there were Surrey top and full soft-top options as well.

During 1995, Land Rover South Africa was also looking at marketing a vehicle similar to the NAS 90, but with a 300Tdi diesel engine. A single prototype was built at Solihull by modifying a production 90; this was easier than converting a NAS 90 chassis to the right-hand drive (RHD) that the South Africans wanted. The project went no further and the prototype (known at Solihull as the 'NASSA 90') was sold into private ownership in the UK.

The Dutch 129in

The Defender 130 in crew cab form had been a strong seller on the commercial market in the Netherlands, but in 1994 Dutch regulations changed, requiring vehicles classified as commercials to have at least 40 per cent of their load area length ahead of the rear axle. This was a serious problem for Land Rover, because the 130 did not comply and would therefore be subject to more than twice as much road tax as before, which would kill sales.

So Land Rover Special Vehicles developed a special model for the Dutch market, adding a 2in (51mm) chassis extension to the 130's 127in (3,226mm) wheelbase to make enough room for a full-size 110 pick-up body instead of the cut-down HCPU type used on standard 130 Crew Cabs. An outrigger was added on each side as a body mounting point and some changes at the back of the cab plus a 2in insert in the sill panel on each side allowed everything to mate up correctly.

Production of this model continued into the early 2000s, when regulations changed and the standard 130 became saleable in the Netherlands once again. Build quantities of the Dutch 129in are not available, but in 2002 LRSV claimed to be manufacturing about one example a week.

This is the special Dutch version of the Defender 130, with two more inches in the wheelbase and a standard pick-up bed instead of the cut-down HCPU type.

The Dutch 'Hi-Roof' 90

The changes in Dutch regulations also created problems for the Defender 90, which had insufficient space behind the seats to qualify as a commercial vehicle. So Land Rover Netherlands went to specialist Ter Berg for a solution. Ter Berg designed a small roof extension to suit the hardtop or Station Wagon body (which was sold without rear seats as a commercial hardtop) and a number of vehicles were converted in this way.

In the beginning, the regulations called for any windows behind the driver to be blacked out, but in the early 2000s the regulations changed again and windows on the passenger side of the vehicle had to be in clear glass so that the driver could more clearly see traffic (particularly bicycles) on that side of the vehicle.

The 1996 Model Year

The 1996 model year was a quiet one for the civilian Defender range. County Station Wagons were made available with a wider range of metallic and micatallic paints, with these paints retaining the marketing names they had in applications elsewhere within the Land Rover range. This was something of a U-turn, as earlier policy had been to limit Defender paint names to a description of their colours for markets outside North America. Perhaps it also reflected Land Rover's growing awareness of the Defender as a lifestyle accessory rather than a purely functional commercial vehicle.

Land Rover France compensated for the lack of excitement in the Defender range by producing two special editions, both of them rather interesting because they had specifications that

To qualify for the lower commercial-vehicle road tax in Holland, 90s had to have a 'high-roof' conversion by Ter Berg.

STANDARD COLOURS AND TRIMS

1995 Model Year

Paints

Blue	LRC424	
Green	LRC419	
Grey	LRC476	
Red	LRC390	
White	LRC456	
County Black	LRC416	Same as Range Rover Beluga Black and Discovery Caracal
County Green	LRC413	Same as Range Rover Ardennes Green and Discovery Carrigada Green
County Red	LRC536	Same as Range Rover Trocadero Red and Discovery Foxfire Red

Paint Colours for the 1995 NAS 90

AA Yellow	LRC559	
Alpine White	LRC456	
Beluga Black	LRC 416	Extra-cost option
Coniston Green	LRC570	
Portofino Red	LRC 390	Same as Arrow Red and Discovery Corallin Red; names from Range Rover palette

Side Stripes

County models had side stripes in two-tone grey with 'Defender' in yellow on each door and 'County' in smaller letters on the rear flanks.

Upholstery and Trim

Grey vinyl upholstery was used on base models, while County derivatives had Moorland grey dog-tooth check cloth upholstery.

were simply not available in other countries where Defenders were sold. Meanwhile, there were no 1996 models for the USA. LRNA believed that new regulations would make the model unsaleable that year, so made no plans to import any. When it became clear that the regulations would not, in fact, prevent the Defender from being sold, it was too late to programme any into the Solihull build schedule.

The lack of side decals helps to identify this 1996 model as a standard Station Wagon – although the Land Rover Press Office claimed that the picture showed a County type, regardless!

RIGHT: **As this picture makes clear, the 1996 County Station Wagons still had the full side-stripe treatment.**

A BMW ENGINE FOR SOUTH AFRICA

Defenders were built in South Africa from CKD kits shipped out from Solihull. The assembly plant at Rosslyn was close to the BMW plant, so during the German company's ownership of Land Rover, it was no surprise to find a BMW engine going into a Defender.

The engine used was the BMW M52, a 6-cylinder 2.8-litre petrol unit originally designed for saloon and sports cars. It offered very good road performance, although low-down torque for off-road use was less impressive. All models came with the R380 5-speed gearbox and 90 versions had a larger fuel tank than was used on the alternative diesel (300 Tdi) models.

Defenders with this engine carried '2.8i' decals on their front wings and were sold only in southern Africa – although around a dozen vehicles (mostly 90s) found their way to the UK through specialist exporters. Models available with the BMW 2.8-litre engine were the 90 County, 110 County, 110 pick-up and 110 hardtop.

Production began in 1996 and ended when Ford bought Land Rover from BMW in 2000. No production figures for the BMW-engined Defenders are currently available.

The South African CKD plant built its own Defender derivatives, using a BMW 6-cylinder engine.

Nevertheless, 1996 did have its excitement for Defender – a lot of it. In January 1996 the news was released that Land Rover had won a major Ministry of Defence vehicle contract, involving the supply of several thousand vehicles over a period ending in 2000. These vehicles were based on the Defender 300Tdi and would consist of 90s, 110s and 130s, all specially redeveloped under a programme known as Project Wolf. There is more about these special Defenders, which were named Defender XD types, in Chapter 11. One effect of the Wolf programme that did become apparent on civilian Defenders was that the special heavy-duty wheels

STANDARD COLOURS AND TRIMS

1996 Model Year

Paints

Alpine White	LRC456	
Arles Blue	LRC424	Same as Windjammer Blue for Discovery
Coniston Green	LRC637	
Portofino Red	LRC390	Also known as Arrow Red; same as Corallin Red for Discovery
Beluga Black	LRC416	Same as Caracal Black for Discovery; clear over-base solid; County only
Biarritz Blue	LRC965	County only
Epsom Green	LRC961	County only
Niagara	LRC574	County only
Rioja Red	LRC601	County only
Willow	LRC970	County only

Side Stripes

County models had side stripes in two-tone grey with 'Defender' in yellow on each door and 'County' in smaller letters on the rear flanks.

Upholstery and Trim

Grey vinyl upholstery was used on base models, while County derivatives had Moorland grey dog-tooth check cloth upholstery.

(recognizable by their perforated steel rims) developed for the XD models would begin to appear on Defender 130 models as well.

The 1997 Model Year

There were no major changes to the range for the 1997 model year, but new upholstery materials were introduced. Base models now had grey Twill vinyl (as pioneered on the NAS 90 for 1994), while County models switched to Raleigh cloth, again in grey. The door casings, cubby box and dash rails were now in Dark Granite and there were Dark Granite carpets for County models.

All Defender 90 Station Wagons now had Avon Rangemaster 750 × 16 tyres as standard and the Rostyle steel wheel option for the short-wheelbase models was discontinued in the face of the Freestyle Choice alloy wheel's popularity. The Freestyle Choice wheels now came with a sparkle finish and 265-section tyres and were always accompanied by anti-roll bars front and rear. The Freestyle Choice option was also extended to the Defender 110 for the first time, although here there were no anti-roll bars and the wheels were stronger dished alloy types that carried Goodyear 235/85R16 tyres. All models with the Freestyle Choice option came with locking wheel nuts as standard. A new option for 110 Heavy-Duty and 130 models was the stronger

Side decals and the Freestyle Choice wheels characterize the 1997 90 County Station Wagon for UK buyers, but the German-specification model of the same year has the base-level trim without side stripes. German buyers liked the Rostyle wheels, so they remained available in that country.

County trim became available on the 90 hardtop as well for 1997, complete with the appropriate side stripes and alloy wheels.

perforated steel disc wheel that had been developed for the military XD models (*see* Chapter 11).

There was one further new option for the County models, which could now be ordered at extra cost with a body-coloured roof instead of the white roof standard on other hardtops and Station Wagons. There were also evolutionary changes in the paint options for all models. In France, some of the new features were used to create a special-edition model based on the Defender 90.

The 110 County Station Wagons took on deep-dish alloy wheels for 1997, together with Rayleigh grey woven cloth upholstery.

The 1997 NAS 90

LRNA returned to the fray for the 1997 model year with another run of Defender 90s, but these would be the last NAS models. New legislation would demand airbags for all new vehicles in the 90's class during the 1998 model year, and Land Rover was unwilling to fund the engineering programme needed to develop an installation. The company's thinking was that the cost would not be justified by the small sales volumes for the USA, and that the Defender was unlikely to remain in production long enough for the cost to be recouped from worldwide sales in later years. At that stage, nobody had any idea that demand for the model would ensure that it would still be in production a decade and a half later!

The 1997 NAS 90s were marketed in two stages. Beginning in October 1997 came a run of 1,300 'hardtops', which were in fact Station Wagons despite the description used in publicity material. These were followed in the spring of 1998 by 1,499 soft-tops, making a total of 2,799 examples for the model year. As before, all sales were in the USA and there were none in Canada.

These 1997 NAS 90s all had the latest 4.0-litre V8 engine, driving through a 4-speed ZF automatic main gearbox and the latest LT230T transfer gearbox with new gears and a stronger casing to reduce noise. This made them the first-ever line-assembled automatic Defenders – but they would not be the last. The 4.0-litre V8 engine actually had the same 3,947cc swept volume as the earlier 3.9-litre type, but had been renamed to differentiate the two; it had been introduced in 1994 for the second-generation Range Rover and had more compact packaging of the front-end ancillaries. Power and torque outputs were almost identical to those of the 3.9-litre V8, with 182bhp at 4,750rpm and 233lb ft at 3,000rpm.

Also common to both soft-top and 'hardtop' models were the Freestyle Choice wheel, tyre and anti-roll bar package with 265/75R16 BF Goodrich All-Terrains, and an integrated Class III tow hitch. All models also had an electronic instrument panel with digital mileage readouts, and the optional

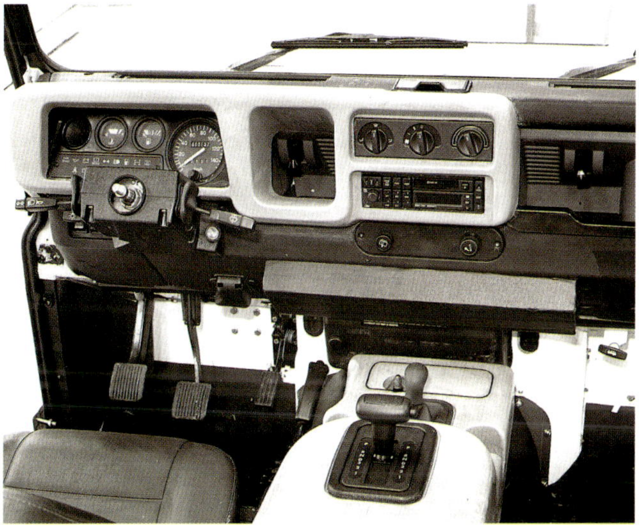

There had been no NAS Defenders for 1996 and Land Rover had taken the opportunity to develop new variants for 1997. Automatic transmission was always on the cards and the more sophisticated dashboard seen here in the styling studios was intended to go with it, although it did not make production. The red soft-top 90 was the development vehicle for the automatic model; it still survives as part of the Dunsfold Collection.

The NAS 90 for 1997 could be bought as a Station Wagon during the colder season, although soft-tops reached the showrooms in time for summer. Both types had the latest 4.0-litre V8 engine with an automatic gearbox.

air conditioning was more powerful than before. When the optional six-CD autochanger was ordered, it was fitted inside the cubby box. Each 1997 NAS 90 was individually numbered on an aluminium plaque mounted to the tail panel. Side runners and a brush bar were now relegated to the options list.

The 'hardtop' models all came with the latest Rayleigh cloth upholstery, six seats, and a redesigned cubby box with integrated cupholders – which were by this stage becoming essential for US market sales. There were four body colour options, a fifth one being added later in the season to create a limited edition that came as standard with a selection of equipment from the Land Rover Kit accessories catalogue. All colours had a white roof and black wheel-arch eyebrows.

Soft-tops supposedly came with better sealing than before, although military-style doors with removable tops were still fitted. The soft-top models still had Charcoal twill-effect upholstery.

STANDARD COLOURS AND TRIMS — 1997 Model Year

Paints

Alpine White	LRC456	
Arles Blue	LRC424	
Coniston Green	LRC570	
Beluga Black	LRC416	Clear over-base solid; County only
Epsom Green	LRC961	Micatallic; County only
Niagara	LRC574	Metallic; County only
Portofino Red	LRC390	
Rioja Red	LRC601	Micatallic; County only
Riviera Blue	LRC588	Metallic; County only
Willow Green	LRC970	Metallic; County only

Roofs were painted white as standard, but could be ordered in the main body colour on County models.

Paint Colours for 1997 NAS 90

AA Yellow	LRC559	Soft-top only
Alpine White	LRC456	Soft-top and hardtop
Arles Blue	LRC424	Hardtop only
Beluga Black	LRC416	Soft-top only
British Racing Green	LRC617	Soft-top only
Coniston Green	LRC570	Hardtop only
Monza Red	LRC590	Soft-top only
Portofino Red	LRC390	Hardtop only
Willow Green	LRC970	Hardtop limited edition only

Side Stripes

County models had side stripes in two-tone grey with 'Defender' in yellow on each door and 'County' in smaller letters on the rear flanks.

Upholstery and Trim

Grey Twill pattern vinyl upholstery was used on base models. County derivatives had Rayleigh grey woven cloth upholstery.

Utility companies remained confirmed customers for commercial Defenders and this 1998 model went to the North of Scotland Water Authority. The badge behind the front wheel makes clear that LRSV had a hand in its preparation.

The 1998 Model Year

To most onlookers, the most important thing to happen to Land Rover during the 1998 model year was the marque's fiftieth anniversary, celebrated from April 1998 onwards as a means of gaining major publicity coverage. However, behind the scenes there were far more important issues at stake. From 1 January 1998, new European Community Whole Vehicle Type Approval (EC-WVTA) legislation would be introduced and this had major implications for the passenger-carrying models within the Defender range.

Before the end of 1997, different countries in the European Community had different types of approval requirements and meeting all these different demands was a major headache for car manufacturers everywhere. The aim of the new EC-WVTA legislation was to introduce an agreed common set of standards across the European Community, thereby making life easier for car makers. Easier it may have been afterwards, but Land Rover had to spend £30 million re-engineering the Defender range to comply with the new

XDS FOR AN EXPEDITION

Two of the XD 110 models built in 1997 and specially prepared for an expedition survived to tell the tale.

Land Rover never had any intention of making the heavy-duty XD military specification (*see* Chapter 11) available on the civilian market, but in 1997 it did build seven XD 110s for an expedition that was intended to coincide with the company's fiftieth anniversary in 1998. This was to be known as the Trans-Global Expedition and was to be led by the renowned explorer Sir Ranulph Fiennes. The aim was to cover 23,000 miles (37,000km), leaving London and travelling through Russia and Alaska to finish in New York. However, Land Rover's owners at BMW cancelled the expedition just a few days before it was due to leave London in 1997.

The expedition vehicles were fitted with standard hardtops and safety devices such as external roll cages with roof racks. All were painted gold. For the crossing of the normally frozen Bering Strait, they were also prepared with special track conversions and custom-made paddle-wheel catamarans made of Kevlar and driven by hydraulic pumps on the vehicles' PTOs.

After the expedition was cancelled, one of the vehicles became a support truck for the 50–50 Challenge charity event that was organized by Land Rover employees during 1998. Christened 'Goldilocks' by the expedition team for fairly obvious reasons, it still survives in the Dunsfold Collection. One more still survives with its tracks intact. The other vehicles, stripped of their special tracks and catamarans, were sold.

standards. Some of the changes were standardized on models for all markets worldwide, while others were confined to Europe.

A key concern that the new European legislation addressed was dangers from sharp edges and protrusions in an accident and, as a result, several exterior features on the Defender had to be amended. Individually, some of these perhaps seemed petty – softening the radius of the rear number-plate light, the rear Land Rover badge, the rear reflector and the tip of the radio aerial were fairly small beer in themselves. So was removing the rivets from the rear mudflaps. However, EC-WVTA deemed such changes necessary and these were among the long list that was applied to 1998-model Defenders for all markets worldwide. It would simply have been uneconomical to have one set of components for Europe and a second for other markets.

Less obviously, all 1998 Defenders had modified door hinges and changes to the seat-belt stalks and the location of the front-belt anchorages, which were moved forwards by 70mm (2.7in) to comply with the new legislation (the change did improve occupant comfort as well). Still on the inside, there were modifications to the air-vent handles and to the lighting and electrical switches to reduce projections; for the same reason, the ICE speakers in the doors were made flush with the door trims. There were new warning light graphics, too, which remained invisible until illuminated.

There were chassis frame stiffeners, although these were again invisible without a very close look and there was a new towing pack, too, known as ORIS. This accommodated all contemporary attachments, including tow ball, towing pintle and NATO-style hook – and included a lashing cleat as well.

For European markets only – and these did include the UK – there were further changes that were confined to the Station Wagons. All had to have end caps on their front bumpers, while side steps could no longer be fitted to 90 and nine-seater 110 Station Wagons because they were classed as a dangerous protrusion; side runners had to be fitted instead. However, the side steps remained available for the ten- and twelve-seater 110 Station Wagons in the UK and Denmark. Station Wagons also had to have a third, high-mounted brake light. Known to Land Rover as a CHMSL (Centre, High-Mounted Stop Lamp), this was always mounted inside the glass at the top of the rear door.

The legislation also looked at GVWs and axle loadings, with the result that both ordinary and County Station Wagon versions of the 90 had to have heavy-duty springs (giving a 2,550kg [5,623lb] GVW), plus ventilated front brake discs. Meanwhile, the 3,500kg-rated (7,718lb) heavy-duty suspension for the Defender 110 Station Wagons no longer met requirements and was withdrawn as an option. The headlamp levelling system formerly fitted only to vehicles for Germany was now standardized for all Station Wagons and was operated by a control switch in the centre of the dash. The dim-dip lighting system was deleted. Then the Station Wagons also had to have the NAS-style steering wheel with its larger centre boss, although other models retained the earlier type.

Lastly, the regulations had an impact on wheels and tyres. They forbade fitting the spare wheel on the bonnet, because it impeded forward visibility. If the wheel was mounted on the tail door, it had to have a cover plate. As for tyre sizes, 205-section types were no longer legal on 90s, which could now only have 7.50 × 16 tyres (on steel wheels), or 265-section tyres (on Freestyle Choice alloy wheels).

With all that to prepare, it is hardly surprising that Land Rover minimized other changes to Defenders for the 1998 model year. Changing the range of paints on offer looked like a simple running change, but it was also associated with the company's move to using water-based clear over-base-coat paint technology (without alkyds) in order to minimize environmental impact.

Then of course there were several special-edition Defenders associated with the fiftieth anniversary. The 1998 model year saw an expansion of sorts here, too, as it was the first time there had been special editions of the commercial models rather than the passenger-focused models. Once again, it was Land Rover France who led the way in this respect. The headline attraction, though, was a very special fiftieth anniversary edition of the 90, which featured the 4.0-litre V8 petrol engine and 4-speed automatic gearbox only seen before in 1997-model Defender 90s for North America.

Volume build of the 1998 models did not begin until 8 December 1997, so only small numbers of 1998 models reached the showrooms before the New Year. The 1998 model year ran right through to the end of the 1998 calendar year.

STANDARD COLOURS AND TRIMS

1998 Model Year

Paints

Atlantis Blue	LRC632	Micatallic; V8 50 only
Beluga Black	LRC416	Clear over-base solid; County models only
Caledonian Blue	LRC507	
Chawton White	LRC603	
Cobar Blue	LRC624	Metallic; County models only
Coniston Green	LRC637	
Epsom Green	LRC961	Micatallic; County models only
Niagara Grey	LRC574	Metallic; County models only
Rioja Red	LRC601	Micatallic; County models only
Rutland Red	LRC607	
Woodcote Green	LRC623	Micatallic; County models only

Roofs were painted white as standard, but could be ordered in the main body colour on County models.

Side Stripes

County models had side stripes in two-tone grey with 'Defender' in yellow on each door and 'County' in smaller letters on the rear flanks.

Upholstery and Trim

Grey Twill pattern vinyl upholstery was used on base models, while County derivatives had Rayleigh grey woven cloth upholstery.

VINS

(i) European-Type VINs

The VIN codes used in this period consisted of seventeen characters, made up of an eleven-character prefix code and a six-digit serial number.

Example: SALLDHAF7MA-xxxxxx

This breaks down as follows:

SAL Manufacturer code (Rover Group)
LD Dcfender
H 110in (2,794mm) wheelbase
    K = 127in (3,226mm) wheelbase (Defender 130)
    V = 92.9in (2,360mm) wheelbase (Defender 90)
A Truck cab, soft-top or hardtop body
    M = Station Wagon
F 2.5-litre 300 Tdi diesel engine (1995–7)
    6 = 2.5-litre 300 Tdi diesel engine (1998 models)
    Y = 2.0-litre 4-cylinder T16 petrol engine
7 RHD with 5-speed manual gearbox
    8 = LHD with 5-speed manual gearbox
M Model year 1995
    T = 1996
    V = 1997
    W = 1998
A Assembled at Solihull
    F = Shipped as KD for overseas assembly
xxxxxx Serial number

(ii) NAS VINs

The VIN codes used in this period consisted of seventeen characters, made up of an eleven-character prefix code and a six-digit serial number.

Example: SALDV2281SA-xxxxxx

This breaks down as follows:

SAL Manufacturer code (Land Rover)
DV Defender 90, except Californian specification
    DX = Defender 90 for California
2 Two-door body (soft-top)
    3 = Three-door body (Station Wagon)
2 3.9-litre V8 petrol engine
    3 = 4.0-litre V8 petrol engine
8 LHD with 5-speed manual gearbox
    4 = LHD with 4-speed automatic gearbox
1 Security check digit (0 to 9, or X)
S Model year 1995
    V = 1997
A Assembled at Solihull
xxxxxx Serial number
    948000–981999 range (1995 models)
    999000–999999 range (1997 models)
    100000–126999 range (1997 models)

Engine Identification Numbers for Defender, 1995–8 models

Engines have a nine-digit identification number. The digits break down into three groups: the first group of three digits identifies the engine type, the second group of five digits is the engine serial number, and the final digit identifies sub-groups created by running specification changes.

Example: 16L-00001A.

2.0-litre Mpi petrol engine

20T4HG91-xxxxxx Italian Carabineri order only

2.5-litre diesel engine

12J-00001A Standard engines
18J-00001A For British MoD vehicles, with 24V electrical system

2.5-litre petrol engine

17H-00001A Standard engines
18H-00001A Military engines with 24V electrical system
19H-00001A With auto gearbox, for unsuccessful Swiss military tender

2.5-litre 300 Tdi diesel engine

15L-00001A Militarized engine, with 24V electrical system
16L-00001A Standard engine

VINS AND ENGINE NUMBERS — DEFENDER 1995–8 Models

3.9-litre V8 injection engine

37D-00001A 8.13:1 compression, with catalytic converter exhaust

4.0-litre V8 engine

30G-00001A NAS and 50th Anniversary Defender 90 automatic, with catalytic converter exhaust

SPECIFICATIONS — DEFENDER 1995–8

Petrol Engines

Type	Mpi
Bore and stroke	84.45mm × 89mm
Capacity	1,994cc
Valves	4-cylinder with 4 valves per cylinder
Compression ratio	10:1
Fuel system	multi-point injection with MEMS 1.6 engine management
Max. power (DIN)	134bhp at 6,000rpm and 140lb ft at 3,600rpm

Type	BMW M52
Bore and stroke	88.9mm × 71.1mm
Capacity	3,528cc
Valves	6-cylinder with twin ohc and 4 valves per cylinder
Compression ratio	10.2:1
Fuel system	Bosch Motronic engine management
Max. power (DIN)	190bhp at 5,300rpm and 206lb ft at 3,950rpm

Type	V8
Bore and stroke	88.9mm × 71.1mm
Capacity	3,528cc
Valves	V8-cylinder with pushrod ohv
Compression ratio	8.13:1
Fuel system	two Zenith-Stromberg carburettors
Max. power (DIN)	135bhp at 5,000rpm and 182lb ft at 2,500rpm

Type	NAS V8, 1995
Bore and stroke	93.98mm × 71.1mm
Capacity	3,947cc
Valves	V8-cylinder with pushrod ohv
Compression ratio	8.13:1
Fuel system	Lucas L-Jetronic fuel-injection
Max. power (DIN)	180bhp at 4,750rpm and 227lb ft at 3,500rpm (110 models); 182bhp at 4,750rpm and 232lb ft at 3,100rpm (90 models).

Type	NAS '4.0-litre' V8, 1997
Bore and stroke	93.98mm × 71.1mm
Capacity	3,947cc
Valves	V8-cylinder with pushrod ohv
Compression ratio	9.35:1
Fuel system	Lucas 'hot-wire' fuel-injection with GEMS engine-management system
Max. power (DIN)	190bhp at 4,750rpm and 236lb ft at 3,000rpm

Diesel Engines

Type	300Tdi
Bore and stroke	90.47mm × 97mm
Capacity	2,495cc
Valves	4-cylinder with ohv
Compression ratio	19.5:1
Carburettor	Direct-injection and Garrett turbocharger
Max. power (DIN)	111bhp at 4,000rpm and 195lb ft at 1,800rpm.

Transmission

Permanent four-wheel drive.

Main gearbox

(a) 1995–8 model years:

5-speed R380 manual gearbox

Ratios 3.692:1, 2.132:1, 1.397:1, 1.000:1, 0.770:1, reverse 3.536:1

Continued overleaf

Continued from previous page

SPECIFICATIONS DEFENDER 1995–8

(b) 1997 NAS 90:
4-speed ZF 4HP22 automatic gearbox
Ratios 2.480:1, 1.480:1, 1.000:1, 0.730:1, reverse 2.090:1

Transfer gearbox
2-speed LT230
Ratios 1.411:1 (high) and 3.320:1 (low)
Axle ratio: 3.54:1

Suspension and Steering

Suspension
Coil springs all round.

Steering
Worm and roller steering with 20.2:1 ratio; optional power assistance with 19.3:1 ratio; PAS standard

Brakes

Disc brakes on all four wheels; some heavy-duty models with ventilated front discs; servo assistance; separate drum-type transmission parking brake.

Electrical System

12V with alternator

Dimensions

90 models

Track	58.5in (1,486mm)
Wheelbase	92.9in (2,360mm)
Overall length	146.5in (3,720mm)
Overall width	70.5in (1,791mm)
Unladen height	77.3in (1,963mm) for truck cab 77.4in (1,966mm) for soft-top 77.6in (1,971mm) for hardtop and Station Wagon
Unladen weights	3,587lb (1,627kg) minimum with V8 petrol engine 3,607lb (1,636kg) minimum with 4-cylinder petrol engine 3,734lb (1,694kg) minimum with Tdi diesel engine 3,913lb (1,775kg) for 1997 NAS 90

110 models

Track	58.5in (1,486mm)
Wheelbase	110in (2,794mm)
Overall length	175in (4,445mm) for truck cab and soft-top 180.3in (4,580mm) for Station Wagon 184in (4,674mm) for HCPU
Overall width	70.5in (1,791mm)
Unladen height	80.1in (2,034mm)
Unladen weights	3,979lb (1,805kg) minimum with 4-cylinder petrol engine 3,981lb (1,806kg) minimum with V8 petrol engine 4,127lb (1,872kg) minimum with Tdi diesel engine

130 models

Track	58.5in (1,486mm)
Wheelbase	127in (3,226mm)
Overall length	198in (5,029mm)
Overall width	70.5in (1,791mm)
Unladen height	80.1in (2,034mm)
Unladen weights	4,436lb (2,012kg) minimum with V8 petrol engine 4,599lb (2,086kg) minimum with Tdi engine

Performance

110 Station Wagon with 134bhp V8 and 5-speed LT85

0–60mph	15sec
Maximum	84mph (135km/h)
Fuel consumption	13–15mpg (21.8–18.9ltr/100km)

90 NAS soft-top with 182bhp V8 and 5-speed LT77S

0–60mph	10.2sec
Maximum	86mph (138km/h)
Fuel consumption	14–15mpg (US gallons)

110 with 300Tdi engine and 5-speed R380

0–60mph	Not quoted
Maximum	Not quoted
Fuel consumption	29mpg (9.8ltr/100km)

CHAPTER EIGHT

ELECTRONICS TAKE OVER, 1999–2006

The Td5 Era – 1999–2006 Model Years

By the mid-1990s, it was clear that the Gemini diesel engine design had reached the limit of its development. Although there was enough metal in the cylinder block to enlarge it to around 3 litres, there were good reasons why Land Rover did not want to go down this route. First of all, two of its key European markets – France and Italy – imposed tax penalties on diesel engines over 2.5 litres, which would inevitably have a major impact on sales. Secondly, the European Community had embarked on a process of reducing diesel exhaust emissions and the Gemini fuel system would need extensive redevelopment in order to remain compliant with forthcoming new regulations.

So development work started on a new diesel engine design. At this stage, Rover Cars and Land Rover engineers were still working closely together and the engine picked up on design features that had been developed for the 2.0-litre L-series diesel engine introduced for the Rover 600 saloon in 1994. Most obviously, the new Land Rover engine used its combustion chamber design; the combustion chamber is the most time-consuming element of engine design, therefore it made sense to build on the successful L-series chamber. One result of this was that the new engine would have the same bore and stroke dimensions as the L-series engine.

The Land Rover strategy was ambitious. Its aim was to deliver a modular engine design with derivatives to suit every model in the Land Rover range. There would be a 2.0-litre 4-cylinder for the Defender, a 2.5-litre 5-cylinder for the Discovery and a 3.0-litre 6-cylinder for the Range Rover, all sharing the same bore and stroke dimensions and with a large percentage of common components. However, before the Storm project had progressed beyond the design stages, the Rover Group was sold to BMW. The German company reviewed all the new product plans during 1994 and instructed Land Rover to stop work on the 4-cylinder and 6-cylinder diesels because it had its own 2.0-litre and 3.0-litre diesels under development. These could be adapted for future Land Rover products. That left the 2.5-litre 5-cylinder version of the engine and permission was granted for work to continue on that under the code name of Project Storm.

The new Td5 embodied more advanced new technology and was also more dependent on electronics than any previous Defender engine. It appeared in the Discovery Series II at the same time, in a higher state of tune.

Like the earlier Gemini engines, the Storm engine had a swept volume of 2,495cc, a single exhaust-driven turbocharger and an air-to-air intercooler. It was designed with an alloy cylinder head on a cast-iron block, with overhead valves. The overhead camshaft was driven by a chain, rather than a rubber belt as on the Gemini engines, and there were hydraulic tappets to eliminate the need for adjusting the valve clearances. An oil cooler with its own thermostat was integrated into the lubrication system, which also had a centrifugal oil filter in addition to its standard paper-element type to trap tiny particles that might otherwise circulate with the oil. It enabled Land Rover to extend the service intervals and so reduce maintenance requirements for the new engine.

However, the major departure from previous Land Rover practice was that the Storm engine depended heavily on electronic control systems. Its Engine Control Module (ECM) had been developed under the code name of Thunder and not only gave more accurate control of fuelling, but was also linked into other electronic control units on the vehicle. To give fine control of fuelling and therefore also of exhaust emissions, the Land Rover engineers had used a single high-pressure Electronic Unit Injector for each cylinder. The fuel system worked at 1,500 bar (22,000psi) and was driven by the camshaft. As each injector had a separate plunger pump, the fuel delivery pressure was developed within the injector itself and such fine fuel atomization was achieved under these high pressures that combustion was complete enough for the engine not to need an exhaust catalyst to meet the new EU2 emissions regulations due on 1 January 1999. Diesel 'knock' was further reduced by using a pilot injector in each cylinder that advanced the flame front at idle, making the Storm engine very much quieter in operation than the Gemini engines.

The Storm engine was announced in September 1998 at the Paris Motor Show with the marketing name of Td5. Its announcement for the new Series II Discovery was made at the same time as its announcement for the Defender. In practice, however, Defender production was delayed slightly to give the Discovery an initial advantage in the market place. The Defender engine was also slightly less powerful than the Discovery version, with 122bhp at 4,000rpm instead of 136bhp at 4,000rpm and the same 221lb ft at 1,950rpm. Discovery engines used with automatic transmission had even more torque, but nobody complained that the new Defender engine was lacking in its outputs. It had a lot more urge and a lot more refinement than the outgoing Gemini engine.

The Td5 engine was used in Defenders for eight years – as long as the 200Tdi and 300Tdi engines combined. Some of Land Rover's more conservative customers were deeply suspicious of it at first because they could not get used to the idea of an odd number of cylinders. The engine's reliance on electronics also made many users wary. They felt, quite rightly, that they would be less capable of attending to breakdowns, with many particularly concerned that such a breakdown might occur in a remote area, many hundreds of miles from help. Land Rovers were, after all, traditionally used for exploration work. Nevertheless, the Td5 went on to earn a reputation for reliability and durability nearly as good as that of the Gemini engines it replaced. Production lasted until 2006, although by mid-2005 the Td5 assembly lines and associated equipment were being offered for sale.

Two years after the Td5 had arrived in Defenders, owners BMW sold Land Rover to Ford. There was widespread speculation about whether Ford would finally bring production to an end, but global sales of the Defender held up very well and Ford saw no reason to withdraw the model. Nor was it prepared to commit to an expensive programme to develop a replacement, putting a much higher priority on developing new and more sophisticated passenger-carrying models. The first fruits of this approach were the Discovery 3 in 2004 and the Range Rover Sport in 2005, closely followed by the Freelander 2 for 2006. Not until mid-2006 and the start of the 2007 model year did any sign of direct Ford influence on Defender become apparent.

In any case, there was no urgent need to pour money into a major engineering programme for Defender. The model continued to sell around 20,000 examples a year globally, which, although not a huge figure by twenty-first-century standards, certainly justified keeping in production a vehicle for which the main development had been amortized many years earlier. These sales figures were maintained partly by a very aggressive programme of special-edition and limited-edition models in Europe. Land Rover exploited the Defender's image for all it was worth throughout the period, targeting both those who really did use their vehicles for off-road adventure and also those who wanted to look as if they did. The urban market was tackled very seriously, too.

The 1999 Model year

The 1999-model Defenders were introduced at the Paris Motor Show in October 1998, but in the UK the new model year began in December 1998, with a start to showroom deliveries in January 1999. In European markets, the Td5 engine took over completely from the 300Tdi, although that

RIGHT: **The County Station Wagons for 1999 came with body-colour roofs as standard. Clear in this picture are the slats between side and indicator lights that characterized the Td5 models.**

LEFT: **Land Rover built some special Defender 90s for the James Bond 007 film, *The World is Not Enough*. The prominent grille seen here turned up again on the SVX concept car in autumn 1999.**

engine remained available for overseas markets that did not demand the same low levels of exhaust emissions. It was also still being used in the Wolf or Defender XD military models, whose assembly at Solihull would not finish until 2000. The 1999 model year was also the last one for Defenders with the 2.5-litre 4-cylinder and 3.5-litre V8 petrol engines, as the final examples were built for overseas customers.

There was more to the 1999 Defenders than just a new engine, of course. Of side stripes there was now no sign and Land Rover had provided all models with new Td5 badges on each front wing, while the headlamp surrounds now had louvres under the lamps themselves. County models now came with body-colour roofs as standard. On the inside, there were two upholstery options, both in Dark Granite. Base models had Twill PVC (also described as 'Stag Grain' PVC), while the County models had a new Techno cloth design. Noise insulation had been tackled by a one-piece rubber floor mat and by installation pads in the dashboard and elsewhere to prevent rattles and squeaks. The instrument panel had a new electronic speedometer with digital mileometer, electronic gauges and new warning lights for low fuel and for water in the fuel.

With the new Td5 engine came an electronic accelerator control, known as Fast Throttle Control. This had been introduced during development to overcome a problem with poor off-boost performance – it boosted fuel delivery at low speeds to improve engine response. However, it had also been used to reduce driveline 'shunt' during gear-changing by reducing fuelling and therefore torque slightly as the changes were made. All the 1999 Defenders also had a new plastic fuel tank in place of the earlier metal one and on 90s that tank was now mounted at the rear; as a result, the fuel filler moved from behind the door to behind the rear wheel arch. On the 90, the new tank held 60ltr (13gal), which was 5.5ltr (1.2gal) more than before, but on 110 and 130 models it had lost 5ltr (1.1gal) of capacity and now held 75ltr (16.5gal) of fuel.

Electronics were also introduced elsewhere. At extra cost, Defenders could now be had with an Anti-lock Braking System (ABS), which came with Electronic Traction Control (ETC). The ETC system worked through the ABS sensors to determine when a wheel or wheels had lost traction. It then made use of the action of the differentials to transfer maximum torque to the other wheels by braking the spinning wheel until traction was restored. The system was not new to Solihull products, as it had been introduced in 1992 on Range Rovers and had also been made available on Discoveries. On Defenders, the locking centre differential was retained as well, so that the driver could anticipate when the vehicle was likely to lose traction and take action to avoid it; ETC was intended as a reactive back-up system for when all else failed.

There was one further use of the new electronic systems that was little publicized but deserves comment. Many of the electronic components on the vehicle were designed to stop working unless their unique electronic signature had been programmed into the vehicle's electronic system. The point of this was to deter theft, which had been a serious problem on earlier Defenders, as they were often stolen and then dismantled so that their major components could be sold on. Without the unique electronic 'signature', it was now pointless to remove parts from a stolen vehicle because they would not function elsewhere. When announcing this to the dealers, Land Rover caused some confusion by describing it as an immobilizer system – which of course it was not.

Also new for the 1999 model year Defenders was a County Pack that could be fitted at extra cost to 90 truck cab models. For the outside, it brought metallic or micatallic paint (but only Cobar Blue, Epsom Green and Rioja Red were available) with wheel-arch eyebrows in the body colour, but the standard Ivory White roof. There were front mudflaps and tinted glass as well. Inside the cab, the seats were upholstered in Techno cloth and there was a stereo radio-cassette unit.

As before, the standard wheels were steel discs, with heavy-duty perforated steel disc types now standard on the 130 models. Tyres were 205/80R16s on pick-up and hardtop 90s, or 7.50R16s on 90 Station Wagons, 110s and 130s. However, very large numbers of 1999 models in Europe were delivered with the Freestyle Choice packages. For the Defender 90, this brought five-spoke Tornado alloy wheels with locking wheel nuts and 265/75R16 tyres, plus anti-roll bars front and rear. For the Defender 110, it brought dished five-spoke alloy wheels on 235/85R16 tyres, again with locking wheel nuts.

Highlight among the 1999 model year's limited editions was the Defender Heritage, available for the UK market only and deliberately intended to recall Land Rovers of the 1940s and 1950s – although the standard leather seats suggested it had a rather different underlying purpose! There is more about this and other 1999 special-editions in Chapter 10.

Australian Peculiarities: The Xtreme

In Australia, the 300Tdi engine initially remained available in the 110 Station Wagon, 130 Chassis Cab and 130 Crew Cab. The Td5 was introduced alongside it in the 110 Station Wagon and 130 Crew Cab and in a special promotional model called the Xtreme that went on sale in March 1999. This was a 110 Station Wagon that was promoted as a model for extreme off-road conditions. It came with deep-dish alloy wheels,

The Australians must have liked the UK publicity picture because they did their own version – but at a steeper angle, of course. This was the Xtreme, with a white roof because it was to basic Station Wagon specification.

all-terrain tyres and a black tubular bull bar as standard, plus 'Xtreme' rear side decals. There was also an optional Expedition Pack, consisting of a winch, driving lights, roof rack and Global Positioning System (GPS) satellite navigation system.

The Xtreme proved a popular model and its sales life was extended to 2004. Records show that there were 137 in the 1999 model year, 123 in 2000, 241 in 2001, 350 in 2002 and forty-two in 2003; to these must be added locally assembled 110s totalling 144 in 2002, 201 in 2003 and twelve in 2004.

STANDARD COLOURS AND TRIMS

1999 and 2000 Model Years

Paints

Caledonian Blue	LRC507	
Chawton White	LRC603	
Cobar Blue	LRC624	Metallic; County models only
Coniston Green	LRC637	
Epsom Green	LRC961	Micatallic; County models only
Java Black	LRC697	Micatallic; County models only
Niagara Grey	LRC574	Metallic; County models only
Rioja Red	LRC601	Micatallic; County models only
Rutland Red	LRC607	
Woodcote Green	LRC623	Micatallic; County models only

Roofs were painted in Ivory White except on County Station Wagons, when they were in the body colour. Wheel-arch eyebrows were in black except on County models, when they were in the body colour. County truck cab pick-ups had Ivory White cab roofs.

Upholstery and Trim

Dark Granite Twill pattern PVC upholstery was used on base models, while County derivatives had Dark Granite Techno cloth upholstery.

The 2000 Model Year

Land Rover told its dealers that the aim of the 2000 model year was to promote the Defender's unmatched flexibility and that it wanted to optimize the leisure-market opportunities as well as maintain a high visibility within the utility-vehicle sector of the market. So there was to be an expanded choice of paint colours ('Ideal for tailoring LE packages to suit local markets', claimed the Marketing Bulletin), and the company expected this to allow special editions to be based on models other than County Station Wagons. In practice, however, the 1999-season colours were carried over until a new range became available in June 2000.

Body-colour wheel-arch eyebrows and roofs became standard on pick-up derivatives (as the truck cab models were increasingly being called), but the white roof remained standard on ordinary Station Wagons to distinguish them from County types. The 130 Crew Cab models were also realigned in the market from September 1999, as prices were reduced and a County specification became available. This was preparing the way for the new Defender 110 Double Cab that made its appearance in May 2000.

The idea of a Double Cab was not new; it was in effect the same thing as a Crew Cab, but on the shorter 110in (2,794mm) wheelbase. What was new was a change in UK tax regulations, which meant that VAT was now reclaimable for commercial users. This was matched by increasing sales in Europe of Japanese 4×4 pick-ups with five-seater cabs; a fashion trend was growing and Land Rover wanted to be part of it. As the company explained to its dealers, the Toyota Hi-Lux had defined the market sector, the 130 Crew Cab was too big to compete, so the 110 Double Cab had been developed for the job.

The 110 Double Cab came with a soft-top for its pick-up bed as standard; this was based on the standard pick-up bed and not on the Hi-Capacity type that was used for the 130 Crew Cab. To emphasize the point, the 130 Crew Cab was now renamed the Defender 130 Hi-Capacity Double Cab, although that name was too much of a mouthful to stick.

Meanwhile, there were also developments for military Defenders. As the Td5 engine had not been militarized, the 300Tdi remained the core military engine. It was already capable of supporting 24V electrical systems. However, to ensure that the military Defender remained attractive in European markets, a modified 300Tdi became available at the end of 1999, fitted with a cooled EGR (Exhaust Gas Recirculation) system and an exhaust catalyst, which made it compatible

LEFT: **The 110 Double Cab was a new model in the production range, but similar vehicles had been produced in some of Land Rover's overseas assembly plants for many years. Land Rover itself had built a V8-engined prototype as long ago as 1992.**

BELOW: **2000-model 90 County Station Wagons came with the Freestyle Choice wheel, tyre and suspension package, plus of course body-colour roof and wheels. When the optional air-conditioning was fitted, the front end gained a longer 'snout'.**

This 2000-model 110 Station Wagon is a standard model with steel wheels but with the optional metallic paint and all-over paint finish.

with the current EU2 emissions requirements. These engines were all made in Brazil by International Motors.

The Defender SVX Concept

Land Rover often claims that its first-ever concept vehicle was the Range Stormer, shown in 2004 as a teaser for the forthcoming Range Rover Sport. In fact, the Range Stormer was not the first; that title more properly belongs to the SVX concept based on a Defender 90 that was displayed at the Frankfurt Motor Show in September 1999.

The origin of the SVX almost certainly lay in the rivalry between Mercedes-Benz and BMW (who owned Land Rover at the time). BMW was keen to demonstrate the strength of its brand and products at a time when Mercedes-Benz sales were slipping and there was no better place to do so than at the two companies' 'home' Motor Show. So in order to reinforce its message, BMW asked Land Rover to come up with a show-stopper for its stand.

The SVX was built by Special Vehicles and, unlike many concept cars, was fully driveable. It was based on a Defender 90 with a galvanized and reinforced chassis frame. Longer coil springs and gas dampers allowed a raised ride-height with 20in alloy wheels on purpose-made off-road tyres. The front wheels had ventilated disc brakes and each axle had a four-pin locking differential; electronics disabled the ETC system when the differentials were locked. The engine was of course the current Td5, but it had been uprated to deliver a claimed 375Nm (276lb ft) of torque, which represented an increase of 25 per cent over standard.

The open body had military-specification half-doors and no windscreen. A rollover cage protected the passenger area and its tubular front elements incorporated a snorkel for deep-water wading. The panels were painted Himalayan Green, a metallic silver-green unique to the vehicle, and silver paint and stainless steel items were used to provide visual contrasts; such contrasts would later appear on special-edition production vehicles. The alloy grille was specially made (although similar grilles, albeit painted black, did appear on the Land Rovers used in the 1999 James Bond film *The World is Not Enough*). There were rock sliders in place of the standard sills, which served to protect the long-range fuel tanks mounted under the body. There was a demountable winch, too, with mounting points both front and rear.

The interior had just two seats and featured chequer-plate flooring, alloy gear knobs and alloy pedals, plus a number of other show-car items. Several of these anticipated features that would became available for production models in the next few years.

Despite the impression these pictures give, the SVX concept car was not a 'bigfoot' model – although it did have larger wheels than standard. It was intended to demonstrate the leisure possibilities, and of course the underlying capabilities, of the Defender range.

The SVX remained a one-off. It was never intended as a production vehicle, but was designed to suggest some of the potential capabilities of the Defender. Nevertheless, the SVX name was revived eight years later for a limited edition at the time of Land Rover's sixtieth anniversary.

The 2001 Model Year

The main thrust of new-model activity for the 2001 model year that opened in autumn 2000 was behind the Freelander, which for 2001 had two new engines. So there were no novelties in the Defender range, which had in any case recently been given a lift by the arrival of the Double Cab models and a new range of paint colours. However, there had been a change behind the scenes under the new Ford ownership of Land Rover which began that summer. Under the Ford scheme of things, the Defender would in future have the internal project code of L316.

Changes were nevertheless made gradually throughout the season, beginning in December 2000 when General Grabber 235/85R16 tyres became standard on 110 Station Wagons. They replaced the BF Goodrich 265/75R16 size formerly fitted because Goodrich would no longer supply their tyres without white lettering on the side walls, a feature that Land Rover did not want.

From January 2001, centre cubby boxes were standardized on all UK models except the twelve-seater 110 Station Wagon and County Station Wagon. A centre seat could nevertheless still be ordered as a no-cost option and was quite regularly

Fully equipped for adventure is this 2001-model 90 Station Wagon. It was one of a small fleet specially prepared by Land Rover Special Vehicles for a media event in Spain called the Defender Challenge.

still fitted on export models. Price adjustments were the news for March, as Land Rover implemented the results of a pan-European pricing review. The aim was to harmonize options prices across EU markets and to ensure that Defender options were priced competitively against similar options from rival manufacturers. One notable result was that the cost of the metallic paint options increased.

The next changes came at the end of June and affected the 110 Station Wagons and their County derivatives. New European legislation now demanded that all forward-facing seats in vehicles designed to carry ten or more passengers must have three-point safety belts. This could not be achieved on the centre front seat, so it was replaced with a centre cubby as standard. The long-standing twelve-seater option disappeared and in future all long-wheelbase Station Wagons would have seats for only nine people.

Special editions continued as a feature of Land Rover's marketing strategy for the Defender. In the UK, the experiment with the X-Tech commercial model had been a success, so a second X-Tech edition was produced in spring 2001. France continued its own special-edition programme with a new Hawaii model for the summer of 2001 and there was a global special edition from June 2001, associated with the release of the *Lara Croft: Tomb Raider* film.

The Defender 147

Another new model made its first appearance in 2001, by courtesy of Land Rover South Africa. This was the Defender 147 with, as its name suggests, a 147in (3,734mm) wheelbase.

The 147 came about when LRSA's Special Vehicle Operations team compared notes with its LRSV opposite numbers at Solihull and discovered a common need for an extra-long Defender. LRSV had fielded a request from theme-park transportation specialists Severn-Lamb, who had been asked by Busch Gardens in Florida, USA, for vehicles capable of carrying around seventeen passengers on a new theme-park ride. The South African requirement was for a tour bus with around twelve seats that could do the job traditionally allocated to the traditional African 'toast rack' type of game-viewing vehicle.

So a plan was drawn up to enlarge a Defender 130 chassis by adding a 20in (508mm) extension section. This made its wheelbase 147in (3,734mm) (the 130 chassis actually had a wheelbase of 127in (3,226mm), so adding 20 more made 147). This size was largely dictated by using modified existing body components so that the tour bus became a six-door Station Wagon; Severn-Lamb would of course design a special 'toast rack' body for Busch Gardens to resemble a traditional game viewer.

The chassis extension section incorporated its own cross-member for additional rigidity and a two-piece rear propshaft accommodated the extra length. Careful use of existing production body parts for the Station Wagon meant that only the roof extension and the centre doors had to be specially manufactured. The first two prototypes were Station Wagons, constructed in South Africa with Td5 engines and R380 gearboxes, although there was provision also for fitting the 300Tdi engine that was still available in South Africa.

The sixteen vehicles for Busch Gardens were ready in time for the opening of the theme park in June 2001. They were all RHD models (to give the authentic African experience, even though the theme park is in LHD America), with Td5 engines and automatic gearboxes. Low range was permanently engaged as a way of keeping speeds down on the ride. The Station Wagons were announced in South Africa just four months later in October. A variety of interior configurations was possible, from a seven-seater with lounging room for passengers up to a thirteen-seater if the centre cubby at the front was replaced by an additional seat.

The original plan had been to build a limited number of 147 Station Wagons for the big tour operators in South Africa. However, the original build projection of twenty-four vehicles had to be revised after one of the prototypes was shown at an outdoor leisure exhibition and attracted more orders and enquiries. Some of these were from overseas, but the 147 was too large to be a viable proposition in most countries. For that reason, it was never homologated for the UK.

Nevertheless, LRSV at Solihull did construct three 147 tour buses to the South African design for use at Land Rover's own factories. As they were never going to leave the factory grounds and be driven on the public highway, the lack of homologation was not a problem. All three buses were speed-limited to the 20mph maximum imposed on all Land Rover's factory roads and all were built as twelve-seaters and finished in the silver associated with the Land Rover Experience centres.

LRSV delivered the first one in February 2003 to the Home of the Legend at Solihull, in anticipation of that centre's public opening the following August. The second followed a year later. The third example was built in 2006 and was initially delivered to the tour centre at the Halewood plant near Liverpool where the Freelander 2 was built alongside Jaguar cars.

The six-door Defender 147 Station Wagon resulted from a requirement for a twelve-seater safari bus in South Africa. Its chassis was also used for special tour buses at the Busch Gardens theme park in Florida, plus three Station Wagons were specially built in the UK for use around Land Rover factory sites.

STANDARD COLOURS AND TRIMS — 2001 Model Year

Paints

Alveston Red	LRC696	
Blenheim Silver	LRC642	(Introduced after the start of the model year)
Bonatti Grey	LRC659	
Caledonian Blue	LRC507	
Chawton White	LRC603	
Coniston Green	LRC637	
Epsom Green	LRC961	Micatallic; County models only
Java Black	LRC697	Micatallic; County models only
Kent Green	LRC647	
Monte Carlo Blue	LRC608	
Oslo Blue	LRC644	
Rutland Red	LRC607	

Roofs were painted in Ivory White except on County Station Wagons and County Double Cabs, when they were in the body colour. Wheel-arch eyebrows were in black except on County models, when they were in the body colour. County truck cab pick-ups had Ivory White cab roofs.

Upholstery and Trim

Dark Granite Twill pattern PVC upholstery was used on base models, while County derivatives had Dark Granite Techno cloth upholstery.

NOTE: These specifications were introduced in June 2000, at the end of the 2000 model year.

The 2002 Model Year

The 2002 model year marked another step-change in the Defender story as a large number of changes were introduced.

Most obvious of the 2002 model year changes was to the interior, where there was a new central facia panel. It carried switches for the electric windows and the seat and windscreen heaters on this County model, which also displays the lower-dash rail fitted with air-conditioning.

Volume production of the 2002 models began in mid-August 2001, although they had been announced some months earlier, in May. Most of the 2002 revisions were actually invisible from the outside, although collectively they made up a quite major package.

New European Community emissions regulations, which came into effect on 1 January 2002, were behind changes to the Td5 engines. The first of the new 'EU3' engines were built on 8 August and all 2002 models had the EU3 engine. The cylinder head, injector pressures and engine harness were all modified, and there was a revised state of tune, with a catalytic converter and a cooled EGR system with its accompanying extra hoses.

The most obvious interior change was the new console mounted in the centre of the dash to accommodate the switchgear and ICE head unit; when no ICE system was fitted, the space was occupied by a coin tray. All models now came with 'radio preparation', though, which meant that they had an aerial, two loudspeakers and the appropriate wiring harness whether or not an ICE system was actually fitted. The standard head unit was a Philips C42 radio-cassette with controls for an optional plug-in CD stacker.

For the very first time, Defenders could be had with electric windows in the front doors, with the switches for these located on the new console panel. The electric windows were part of the County specification, but were otherwise an option. There were changes elsewhere on the dashboard, too. The footwell air vents had been redesigned and the heater control cables benefited from reduced friction and improved lever pivots. The fresh-air vent controls, too, had been relocated and now had a positive ratchet mechanism.

On the 90 County Station Wagon (but not the standard 90 Station Wagon), the bulkhead had now been deleted and internal bracing maintained body strength. This was a very welcome change that allowed the front seats a greater range of fore-and-aft adjustment. With a centre cubby box as standard, the 90 County was now a six-seater. The 110 Station Wagons now came with four inward-facing rear seats on two benches, but County models changed to have four individual inward-facing seats after the first few 2002 models had been built. Both were of course nine-seaters. There was a new Cold Climate Pack, too, which for extra cost brought front seat heaters and a heated windscreen.

This 2002 90 County Station Wagon has the five-spoke alloy wheels that replaced the Tornado five-spokes originally associated with the Freestyle Choice package.

One aim of the development programme leading to the 2002 model year vehicles had been to improve corrosion protection; to that end all the doors were now made of zinc-plated steel rather than the traditional Birmabright aluminium alloy. The steel allowed a better fit and finish than the alloy, while the zinc plating resisted corrosion very well. The doors looked the same as before, but they now fitted much better and reduced the water leaks that had for so long been characteristic of the Land Rover utilities.

A closer look revealed that the tail door on Station Wagon and hardtop models had been completely redesigned. Though it looked much the same as before, it was now a single-piece pressing on a reinforced steel frame of galvanized steel and had a continuous rubber window seal. The circuit for the high-mounted brake light was now on the window glass, thereby dispensing with the need for a separate power feed. There was also a new inner trim panel and, on the outside, a new spare wheel mounting.

STANDARD COLOURS AND TRIMS

2002 Model Year

Paints

Alveston Red	LRC696	Micatallic
Bonatti Grey	LRC659	Metallic
Caledonian Blue	LRC507	
Chawton White	LRC603	
Coniston Green	LRC637	
Epsom Green	LRC961	Micatallic; County models only
Java Black	LRC697	Micatallic; County models only
Monte Carlo Blue	LRC608	Micatallic
Oslo Blue	LRC644	Micatallic
Rutland Red	LRC607	
Zambezi Silver	LRC737	Metallic

Roofs were painted in Ivory White except on County Station Wagons and County Double Cabs, when they were in the body colour. Wheel-arch eyebrows were in black except on County models, when they were in the body colour. County truck cab pick-ups had Ivory White cab roofs.

Upholstery and Trim

Dark Granite Twill pattern PVC upholstery was used on base models, while County derivatives had Dark Granite Techno cloth upholstery.

Additional equipment changes involved the windscreen wipers, which now had smoother operation and were, claimed Land Rover, more efficient. There was now a single key for ignition and doors instead of two separate keys and a completely new feature for Defenders was central locking with a remote control. This latter was standard with the County specification, but optional elsewhere. There were chassis changes, too. There were new fabricated Panhard rods with larger bushes, upgraded steel for the support brackets and a higher torque for the fixing bolts. On 110 and 130 models, new rear brake callipers made room for thicker pads, which were intended to increase pad life in harsh operating environments.

Options changed, too. For 2002, the ABS-plus-ETC package was standardized in most markets. The Freestyle Choice package now came with Boost five-spoke alloy wheels for both 90 and 110 models, in each case with General Grabber 235/85R16 tyres. From November 2001, the County Pack could be had on all models, but the bulkhead deletion on 90s was still only available for County Station Wagons; hardtops remained as they always had been. There were some changes to the paint options as the new season began in October, with the possibilities becoming slightly less confusing as Bonatti Grey now always came with a body-colour roof except on 130 derivatives.

Later in the season, beginning in March 2002, the tool roll and tools were deleted from the standard specification of all Defenders – except for those fitted with Heavy Duty Suspension. Presumably, the thinking here was that working vehicles that needed such suspension were more likely to be driven by hands-on owners who would attempt their own repairs and maintenance!

There was the usual array of special editions during the 2002 model year. In the UK, Scottish dealers had their own 'commercial' special edition in November 2001 and from June 2002 there was the Defender Black. Land Rover France kicked off with the Arusha in the autumn of 2001 and followed up with a cabriolet edition called Sun Rider for summer 2002. Chapter 10 has more details of these models.

The 2003 Model Year

Land Rover had been exploring the sales potential of high-specification Defenders through its special-edition models, and their success had pointed the way forwards for the brand. So for the 2003 model year, a new flagship model was added above the County specification level. In the UK, it was first seen at the Birmingham NEC Motor Show in October 2002, and was called the XS. In Europe, where the two lower specification levels were already known as 'E' and 'S' types, it was known as an SE.

The XS specification was available on hardtop and Station Wagon derivatives of the 90 and on hardtop, Double Cab and Station Wagon derivatives of the 110. From the outside, XS models were readily recognizable by a Brunel silver finish for

Despite the impression Land Rover publicity tended to give, not every new Defender was a County Station Wagon. This 2003 model had an export-only specification of 300Tdi engine and window hardtop body. The Special Vehicles decal behind the front wheel arch shows that it had been specially prepared at Solihull. The UK registration was temporary.

New for the 2003 model year was a trim level above that of the County. In the UK, it was called the XS, but on the European continent it was an SE. These pictures show a 110 Station Wagon, with the Brunel silver grille and dashboard switch panel, and part-leather seats.

the grille and headlamp surround panels. Side runners were also part of the specification, as were smoked lenses for the front and side repeater indicators. The Freestyle Choice wheel, tyre and (on 90s) anti-roll bar package was also standard.

Inside, the key special feature was black upholstery in Ripstop cloth with leather side bolsters, matched by black leather coverings for the steering wheel and gear-shift grips. The central switch panel was in Brunel silver, air-conditioning was standard, and so was a Cold Climate Pack of heated front seats and windscreen. Those models that had a tail door had a mesh storage pocket mounted on its inside, and all the XS models came with central locking, electric windows and the ABS-plus-ETC package as standard.

Changes elsewhere for the 2003 model year were minimal. The confusing paint options were rationalized at last, with white roofs now becoming standard with all 'solid' colours and on 110 County hardtops, regardless of paint. Utility models now came with rubber floor mats, while all others had black carpet. The centre cubby box now came with two cup-holder recesses and all cargo compartments were fitted with load-retention rings.

Land Rover France continued with its successful special-edition programme. For the late summer of 2002 it had the Technium edition and for the summer of 2003 the Highlander. Both were based on the Defender 90.

The South African CKD plant continued to exploit the Defender's possibilities. After a number of accidents involving safari-tour vehicles, the South African Government decided to regulate the tour-bus industry. With effect from late 2003, all new vehicles would have to pass strict homologation tests. Determined to be first in the market with a vehicle that met the new regulations, Land Rover South Africa developed the 130 Game Viewer through its Special Vehicle Operations division.

This had the 300Tdi engine (tour operators were nervous of the Td5 engine's electronic controls) and its body followed the classic 'toast-rack' design with room for ten passengers in addition to the driver. Nine were accommodated in three rows of three forward-facing seats in the 'stadium' style, while the tenth sat next to the driver. The 130 Game Viewer went on sale in 2003 and was available in Mala Mala Green or Sandstone Beige, both colours special to South Africa.

South Africa was also the source of a new model for Australia. Australia had never had the Ninety or Defender 90, focusing entirely on long-wheelbase models after 1980, but in 2003 Land Rover believed the time was right to try again. So at the Melbourne Motor Show in March 2003, the company showed a Defender 90, wearing the same Xtreme decals as the 1999 Australian 110 special edition. The vehicles were imported from South Africa because this cost less than importing from Solihull. Before homologation for Australia, they also had to be modified with the exhaust emerging on the left rather than the right as it normally did.

Despite Land Rover Australia's best attempts to keep prices down, the high cost of importing these 90s meant that they could not be priced significantly lower than the existing 110 models. As a result, they lost out badly to competitive vehicles. Land Rover Australia did reduce the price towards the end of the year, but sales remained poor. So the 90 was withdrawn from Australia and did not reappear until 2010.

The 90 hardtop was still a commercial favourite and this one has the SE specification available in Europe, equating to the XS at home.

ABOVE: **Australia began importing South African-built Defender 90 Station Wagons for the 2003 model year, equipping them to local Xtreme specification and relocating their exhaust tailpipes to suit local regulations.**

LEFT: **The South African Defender 130 Game Viewer was another locally built model designed to meet that country's market requirements.**

Just 189 were sold in the 2003 model year, plus a further twelve in the 2004 season. The grand total of 200 was made up by nine vehicles imported during 2002 for pre-sales evaluation.

The 2004 Model Year

The Defender range was moved discreetly out of the limelight for the 2004 model year, because Land Rover focused all its promotional energies on the new Discovery 3 (or LR3, as it was known in some countries). However, there were still some changes and these were revealed at the Frankfurt Motor Show in September 2003. There was also a global limited edition right at the start of the model year, associated with the G4 Challenge sports adventure event.

All models for 2004 now had black door trims. On the European continent, E (base) and S (County) levels took on a new centre console tray between the front seats, while the S and SE (XS in the UK) levels were distinguished by body-colour wheel-arch eyebrows. Outside Europe, the 300Tdi engine remained available, but the sole European power unit was still the Td5 diesel.

There was a minor diversion in November 2003 when Land Rover North America showed what they called a CKD Defender at the SEMA tuning and accessories show in Los Angeles. It was a 110 model and the showing appears to

STANDARD COLOURS AND TRIMS

2003 and 2004 Model Year

Paints

Alveston Red	LRC696	Micatallic
Belize Green	LRC756	
Bonatti Grey	LRC659	Metallic
Caledonian Blue	LRC507	
Chawton White	LRC603	
Epsom Green	LRC961	Micatallic; County models only
Java Black	LRC697	Micatallic; County models only
Monte Carlo Blue	LRC608	Micatallic
Oslo Blue	LRC644	Micatallic
Rutland Red	LRC607	
Zambezi Silver	LRC737	Metallic

Roofs were painted in Ivory White except on County Station Wagons and County Double Cabs, when they were in the body colour. Wheel-arch eyebrows were in black except on County models, when they were in the body colour. County truck cab pick-ups and 110 County hardtops had Ivory White cab roofs.

Upholstery and Trim

Dark Granite Twill pattern PVC upholstery was used on base models, while County derivatives had Dark Granite Techno cloth upholstery. XS derivatives had Black Ripstop fabric upholstery with leather-covered bolsters.

have resulted from an idea within LRNA that the company could bypass US regulations which the current Defender could not meet by bringing the vehicles in as kits of parts and assembling them in the USA. The idea did not go any further.

Brazilian Specials

The Defenders assembled in Brazil broadly followed the UK pattern, although there were some special paint colours to suit local tastes, and Brazilian Defenders always had the Land

This Brazilian concept could have entered production, but was abandoned when the São Paulo plant closed down.

Rover name in large letters on the leading edge of the bonnet (as would UK-built models from the 2007 model year).

The Brazilian plant also had its own Special Vehicles Team and in 2004 this team put together an interesting Defender 90 concept called the Defender Flame. Essentially designed as a bright and stylish leisure-market model, it had input from Sean Henstridge in the Solihull Design Studio and was finished in an eye-catching polychromatic paint called Spectral Orange. The vehicle was actually based on a UK-built Td5-engined model. Although Land Rover Brasil had ideas about marketing the special front end and other items as an optional add-on, the company's plans came to nothing when Ford decided to close the Brazilian assembly plant at the end of 2005.

The 2005 Model Year

As far as mainstream Defenders were concerned, the 2005 model year was another quiet one. Land Rover's major promotional activity was focused on the new Range Rover Sport.

Nevertheless, there were several limited editions in Europe. France had the Libyan Sand in autumn 2004 and Germany had the Experience. Belgium and Holland shared the Silva at the start of 2005, France had the Black Silver from March, and Belgium and Luxembourg shared the Bel Air cabriolet over the summer. A special edition announced in Holland in July may in fact have been no more than a publicity exercise, but it confirmed the suggestion raised with the Silva edition that Land Rover was increasingly looking at co-branding exercises for promotional purposes.

There were also important changes affecting Land Rover's overseas assembly operations. Both the Brazilian and South African assembly operations closed down during 2005. The South African operation had already relinquished its Chief Engineer to Solihull, however. Gary Taylor, who had started his career at Solihull, had returned to the UK to become Chief Programme Engineer for future Defenders.

The 2006 Model Year

The 2006 model year was the last for the Td5 engine in Defenders, so all the important changes to the specifications were saved for the 2007 model year. The media was given long-lead warning of this in May 2006.

Nevertheless, European countries had no fewer than seven special editions and there is more detail on these in Chapter 10. The UK had the Defender Silver and Germany the Defender Limited at the start of the new model year. Then in the spring of 2006, Switzerland had the Defender Black Edition while France had the Defender G4 Challenge, the only country to capitalize on the second G4 Challenge event through a special-edition model. Germany followed up with three more special editions, called Experience, Style and Cabriolet Two.

STANDARD COLOURS AND TRIMS — 2005 and 2006 Model Years

Paints

Alveston Red	LRC696	Micatallic
Belize Green	LRC756	
Bonatti Grey	LRC659	Metallic
Cairns Blue	LRC849	Micatallic; County models only
Chawton White	LRC603	
Java Black	LRC697	Micatallic; County models only
Tonga Green	LRC904	Micatallic; County models only
Zambezi Silver	LRC737	Metallic

Roofs were painted in Ivory White except on County Station Wagons and County Double Cabs, when they were in the body colour. Wheelarch eyebrows were in black except on County models, when they were in the body colour. County truck cab pick-ups and 110 County hardtops had Ivory White cab roofs.

Upholstery and Trim

Dark Granite Twill pattern PVC upholstery was used on base models, while County derivatives had Dark Granite Techno cloth upholstery. XS derivatives had Black Ripstop fabric upholstery with leather-covered bolsters.

LEFT: **Silver was fast becoming a favourite colour for cars of all kinds and the Defender was no exception. This 2005-model 90 Station Wagon was pictured at Eastnor, where Land Rover does much of its off-road testing. It is towing a trailer to make clear that it was no less a workhorse despite the continuing trend towards comfort and sophistication.**

RIGHT: **The Defender 110 hardtop was still in demand from utility companies and this 2005 model went to Yorkshire Electricity Distribution Ltd (YEDL). It has the heavy-duty perforated steel wheels originally developed for the XD military versions of the Defender during the 1990s.**

BELOW: **The truck-cab 90 was rapidly becoming less and less common, although it looked just as good as it always had. Enthusiast Michael Harnett ordered this 2005 model in Eastnor Green with the base specification of white roof and white steel wheels. He managed to secure a very appropriate registration number for it as well.**

RIGHT: **The National Grid chose to fit aftermarket heavy-duty wheels to this 2006 Defender 110 hardtop, which would have been one of the last with the Td5 engine. The body has been specially modified with a roller shutter above the rear wheel arch.**

BELOW: **Land Rover supports a number of environmental and wildlife conservation organizations around the world, among which is Biosphere Expeditions. This 2006 Defender 110 Station Wagon was specially prepared for them, with expedition roof rack, extra lighting, a snorkel, front winch and light guards.**

An Armoured Defender

In the mid-1990s, Land Rover South Africa envisaged a worldwide market for an armoured Defender that looked like a standard vehicle rather than an armoured car. The company began work on what it called Project Taurus in September 1998. The aim was to design a vehicle that could be built on the main Defender assembly lines in South Africa, using a standard heavy-duty 110 chassis with its 3.5-tonne GVW.

LRSA worked with the Mechanology Design Bureau in Johannesburg on the body, which was designed as a monocoque 'tub'. Its construction was contracted to TRM in Olifantsfontein. Tests on prototypes demonstrated that the Taurus vehicle was able to provide the required level of protection, so the fifth and final prototype was shown at the Eurosatory military exhibition in Paris in 2001.

However, orders were slow to come in. LRSA concluded that the market it had foreseen was no longer there and cancelled work on the project in mid-2002. The fifth prototype found its way into preservation in the Dunsfold Collection and in 2005 the LRSA assembly lines on which the vehicle was to have been built were closed down.

Nevertheless, it appears that the design was passed on to another South African company and that the Taurus did go into small-volume production. Examples have been reported in the hands of companies in the diamond industry and of others who need a discreetly armoured vehicle.

The South African Taurus certainly resembled a standard 110 in outline. This was one of the prototypes.

VINS 1999–2006 Models

There was a reorganization of some parts of the numbering system after Ford took over in 2000. CKD markets gained their own individual codes, for example. However, it is not clear exactly when the new system took effect.

(i) European-Type VINs

The VIN codes used in this period consisted of seventeen characters, made up of an eleven-character prefix code and a six-digit serial number.

Example: SALLDI IA87XA-xxxxxx

This breaks down as follows:

SAL Manufacturer code (Land Rover)
LD Defender
H 110in (2,794mm) wheelbase (also used for 147 models converted from 110s)
 A = Defender 90XD
 B = Defender 110XD
 C = Defender 130XD
 K = 127in (3,226mm) wheelbase (Defender 130)
 L = 147in (3,734mm) wheelbase (South Africa)
 V = 92.9in (2,360mm) wheelbase (Defender 90)
A Truck cab, soft-top or hardtop body
 B = 90 Station Wagon
 F = Crew Cab with standard pick-up body (i.e. 110 Double Cab)
 H = HCPU body with either truck cab or crew cab
 J = Six-door Station Wagon (Defender 147)
 M = 110 Station Wagon
 P = Game Viewer (Defender 130, South Africa)
I 2.5-litre Td5 engine to EU1 specification
 5 = 2.5-litre Td5 diesel engine to EU3 specification

Continued overleaf

Continued from previous page

VINS 1999–2006 Models

6 = 2.5-litre 300Tdi diesel engine with EGR and/or catalyst (special order and South African assembly only)
7 = Td5 for all countries except those designated by codes 8 and 9
8 = 2.5-litre Td5 diesel engine, UK version
9 = 2.5-litre Td5 diesel for Australia, EEC and Japan
E = 2.5-litre 300Tdi without EGR or catalyst, built by Maxxion (International Motors Ltd)
F = 2.5-litre 300Tdi without EGR or catalyst
K = 2.5-litre 300Tdi with cooled EGR and/or catalyst
M = 4.0-litre petrol V8

3	RHD with 4-speed automatic gearbox 4 = LHD with 4-speed automatic gearbox 7 = RHD with 5-speed manual gearbox 8 = LHD with 5-speed manual gearbox
X	Model year 1999 Y = 2000 2 = 2002 4 = 2004 6 = 2006 1 = 2001 3 = 2003 5 = 2005
A	Assembled at Solihull B = Military vehicle assembled at Solihull (2006 MY only) C = Assembled in Zimbabwe F = Shipped as KD for overseas assembly in unspecified country J = Assembled in Malaysia K = Assembled in Kenya N = Assembled in Morocco P = Assembled in Pakistan R = Assembled in Jordan V = Assembled in South Africa W = Assembled in Turkey
xxxxxx	Serial number

(ii) NAS-Type VINs (Used in Japan and Certain Other Territories)

The VIN codes used in this period consisted of seventeen characters, made up of an eleven-character prefix code and a six-digit serial number.

Example: SALDV2231XA-xxxxxx

This breaks down as follows:

SAL	Manufacturer code (Land Rover)
D	Defender
V	92.7in (2,355mm) wheelbase (i.e. Defender 90) H = 110in (2,794mm) wheelbase
2	Two-door body (soft-top) B = Three-door body (90 Station Wagon) M = Five-door body (110 Station Wagon)
2	4.0-litre V8 petrol engine
3	RHD with 4-speed automatic gearbox 4 = LHD with 4-speed automatic gearbox 7 = RHD with 5-speed manual gearbox 8 = LHD with 5-speed manual gearbox
1	Security check digit (0 to 9, or X)
X	Model year 1999 Y = 2000 2 = 2002 4 = 2004 6 = 2006 1 = 2001 3 = 2003 5 = 2005
A	Assembled at Solihull
xxxxxx	Serial number

(iii) Defenders Assembled in Brazil

The Brazilian factory also used seventeen-character VIN codes, made up of an eleven-character prefix code and a six-digit serial number.

Example: 93RLDHA881T-xxxxxx

This breaks down as follows:

93R	Manufacturer code (Land Rover Brazil)
LD	Defender

VINS 1999–2006 Models

H	110in (2,794mm) wheelbase K = 127in (3,226mm) wheelbase (Defender 130) V = 92.9in (2,360mm) wheelbase (Defender 90)
A	Truck cab, soft-top or hardtop body B = 90 Station Wagon F = Crew Cab with standard pick-up body (i.e. 110 Double Cab) H = Crew Cab with HCPU body (i.e. 130 Crew Cab) M = 110 Station Wagon N = Five-door, five-seater utility
8	2.5-litre 300Tdi diesel engine without EGR or catalyst, built by Maxxion (International Motors Ltd)
8	LHD with 5-speed manual gearbox 4 = LHD with 4-speed automatic gearbox
1	Model year 2001 2 = 2002 4 = 2004 6 = 2006 3 = 2003 5 = 2005
T	Assembled in Brazil
xxxxxx	Serial number

Engine Identification Numbers for Defender, 1999–2006 models

Engines have a nine-digit identification number. The digits break down into three groups.

On Tdi diesel engines, the first group of three digits is the engine type, the second group of five digits is the engine serial number and the final digit identifies subgroups created by running specification changes. On Td5 diesel engines, the first group of three digits identifies the engine type, the fourth digit identifies subgroups created by running specification changes and the final group of five digits is the engine serial number.

Examples: 10P-A00001 (Td5)
23L-00001A (Tdi)

2.5-litre Td5 diesel engine

10P-Axxxxx	Standard production engine
12P-Axxxxx	Service engine, stripped
13P-Axxxxx	Service engine, stripped, supplied without cylinder head
14P-Axxxxx	Service 'short' engine

2.5-litre Tdi diesel engine

23L-xxxxxA Brazilian-built engines with electronic EGR system

SPECIFICATIONS DEFENDER 1999–2006

Diesel Engines

Type	300Tdi
Bore and stroke	90.47mm × 97mm
Capacity	2,495cc
Valves	4-cylinder with ohv
Compression ratio	19.5:1
Carburettor	direct-injection and Garrett turbocharger
Max. power (DIN)	111bhp at 4,000rpm and 195lb ft at 1,800rpm

Type	Td5
Bore and stroke	85.5mm × 89mm
Capacity	2,495cc
Valves	5-cylinder with ohv
Compression ratio	19.5:1
Carburettor	direct-injection by electronic unit injectors and Garrett GT2052 turbocharger
Max. power (DIN)	122bhp at 4,000rpm and 221lb ft at 1,950rpm.

Continued overleaf

Continued from previous page

SPECIFICATIONS DEFENDER 1999–2006

Transmission
Permanent four-wheel drive.

Main gearbox
5-speed R380 manual gearbox
Ratios 3.692:1, 2.132:1, 1.397:1, 1.000:1, 0.770:1, reverse 3.536:1

Transfer gearbox
2-speed LT230
Ratios 1.411:1 (high) and 3.320:1 (low)
Axle ratio: 3.54:1

Suspension and Steering:
Suspension
Coil springs all round

Steering
Worm and roller steering with 20.2:1 ratio; optional power assistance with 19.3:1 ratio; PAS standard

Brakes
Type Disc brakes on all four wheels; some heavy-duty models with ventilated front discs; servo assistance; separate drum-type transmission parking brake

Electrical System
12V with alternator

Dimensions
90 models

Track	58.5in (1,486mm)
Wheelbase	92.9in (2,360mm)
Overall length	146.5in (3,720mm)
Overall width	70.5in (1,791mm)
Unladen height	77.3in (1,963mm) for truck cab 77.4in (1,966mm) for soft-top 77.6in (1,971mm) for hardtop and Station Wagon
Unladen weights	3,734lb (1,694kg) minimum with Tdi diesel engine 3,737lb (1,695kg) minimum with Td5 engine

110 models

Track	58.5in (1,486mm)
Wheelbase	110in (2,794mm)
Overall length	175in (4,445mm) for truck cab and soft-top 180.3in (4,580mm) for Station *Wagon* 184in (4,674mm) for HCPU
Overall width	70.5in (1,791mm)
Unladen height	80.1in (2,034mm)
Unladen weights	4,127lb (1,872kg) minimum with Tdi diesel engine 4,156lb (1,885kg) minimum with Td5 engine

130 models

Track	58.5in (1,486mm)
Wheelbase	127in (3,226mm)
Overall length	198in (5,029mm)
Overall width	70.5in (1,791mm)
Unladen height	80.1in (2,034mm)
Unladen weight	4,599lb (2,086kg) minimum with Tdi or Td5 engine

Performance
110 with Td5 engine and 5-speed R380

0–60mph	Not quoted
Maximum	Not quoted
Fuel consumption	27mpg (10.5ltr/100km)

CHAPTER NINE

THE PUMA ENGINES, 2007 ON

The 2.4-litre (2007–11) and 2.2-litre (2012 and Later) Types

Ford's top priority after taking over the reins at Solihull in 2000 was to develop a range of exciting new passenger-carrying models, or SUVs as they were increasingly becoming known. So although there was no shortage of Defender 'product actions' (as Land Rover called them) for the first few years of the new decade, it was mid-2004 before work began on a major upgrade for Defenders. The designers and engineers were given rather less than two years to do what was needed, as the new vehicles had to be ready for production by mid-2006 and there would have to be a period of pilot-build testing before the new models were released.

Under Ford, new vehicle programmes at Land Rover were given an alphanumeric code, and the next iteration of Defender remained L316. There were several programme targets: Land Rover aimed to reduce manufacturing costs by using readily available, volume-produced Ford drivetrain components; it aimed to improve refinement; and it aimed to meet forthcoming legislation in Europe about exhaust emissions and passenger seating.

The aim of reducing manufacturing costs automatically ruled out any possibility that Land Rover should develop its own new engine, so the centrepiece of the new drivetrain was a Ford engine that was already earning a good reputation in that company's Transit vans. This was the 2.4-litre Puma 4-cylinder turbocharged diesel. It had already been engineered to meet the EU4 exhaust emissions regulations that were to be introduced in January 2005 and Ford had work in hand to make sure it would also meet the tighter EU5 regulations expected in September 2009.

The engine needed some work, mainly to waterproof it to Land Rover standards, but its basic configuration was retained

The Puma engine would not fit under the existing Defender bonnet. This was one of the proposals from the Design Studio.

A redesigned dashboard was long overdue and several proposals were considered. This one was by Land Rover interior designer, Memo Ozoturk.

without change. The team working under Chief Programme Engineer Gary Taylor would have liked to modify the sump as well so that the tall Puma engine fitted neatly within the Defender's engine bay. However, this was ruled out on cost grounds. As a result, the bonnet had to be redesigned with a bulge to clear the top of the new engine.

The Puma engine was just as powerful as the Td5 it was destined to replace, despite its smaller swept volume of 2,402cc (actually just 93cc less). For the Defender, it was tuned to deliver the 122bhp of the outgoing 5-cylinder engine at a lower crankshaft speed, which reduced both fuel consumption and noise levels at a stroke. The engine also put out a lot more torque than the Td5, which was another very welcome asset in a heavy 4×4 vehicle often used for towing.

These high outputs were partly attributable to a 4-valve layout, to a common-rail direct-injection system with solenoid-driven fuel injectors and to a variable-geometry Garrett turbocharger that improved accelerator response. That response was further enhanced by individual 'pedal maps' in the electronic control system, specifically matched to the demands of each gear. In addition, an anti-stall feature made it almost impossible to stall the engine. However, there was one bizarre and rather annoying feature: the engine management unit was set to give a maximum speed of just 82mph (132km/h), when the vehicle was clearly good for quite a lot more.

Again with a view to reducing manufacturing costs, Ford asked the Land Rover team to replace the R380 5-speed gearbox (used only in the Defender) with the company's MT82 6-speed type, developed in tandem with German transmission specialists Getrag. The MT82 was already in production, was also used in the Transit van and again had a good reputation. Its six gears gave a much wider spread of ratios than before, the sixth gear itself being set 20 per cent higher than the R380's highest (fifth) gear. So the only new Land Rover component in the driveline was a 2-speed transfer gearbox, drawn up with gearing to suit the Ford engine and main gearbox.

There were to be no major changes to the body design, but a team under Dave Saddington in the Design Studio delivered both the new bonnet and an extensively updated interior. The design brief had included improving the heating and ventilating system, which had always been a weakness on the 'workhorse' Land Rover, integrating an air-conditioning

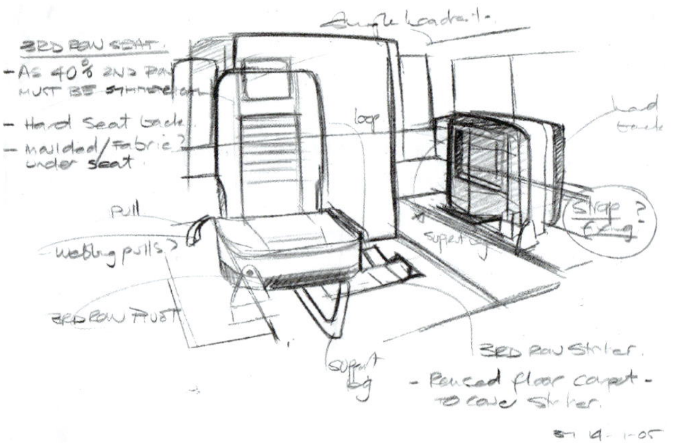

ABOVE: **Dated January 2005, this is a design proposal for the forward-facing rear seats made necessary by new legislation.**

LEFT: **There was a surprising amount of the Puma engine visible under the bonnet at a time when manufacturers tended to hide their engines under plastic vanity covers.**

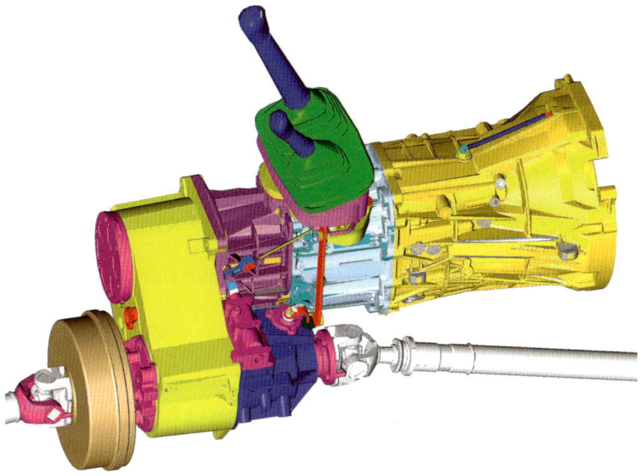

From the Ford stable came a six-speed gearbox. Land Rover designed the 2-speed transfer gearbox bolted to it.

system into it and designing rear seating for new European safety regulations due in early 2007 that would demand all seats to be forward-facing and to have three-point belts.

Saddington's team came up with a completely new dashboard design. Most obvious were the pod-like vents on top of the centre stack, which improved cabin airflow by almost 50 per cent and did away with the need for the traditional opening vents on the bulkhead. Nevertheless, the indentations for these vents remained visible on the outside, mainly because it was cheaper to do this than to press a new panel. The new heater also warmed up faster than the old and could reach a higher temperature, while the air-conditioning cooled faster and to a lower temperature. Side-window demisters were also added, to eliminate an age-old Defender problem.

The new dashboard incorporated an instrument panel that shared its dials with the Discovery 3 and now featured LED illumination. But perhaps its best feature was only noticeable when driving one of the new Defenders. The new dash had been engineered as a one-piece moulding on a steel armature, and thus eliminated the squeaks and creaks so closely identified with older models with their dashboards composed of multiple plastic sections. However, there were still no airbags. The official line was that Land Rover did consider fitting them, 'but concluded it would involve an unfeasible re-engineering of the front structure'.

The new European regulations about forward-facing seats led to one of the most major changes since the Land Rover had first appeared in 1948. The rearmost seats had always faced inwards, because that way it was possible to squeeze more people into them. That those passengers might also be desperately uncomfortable over long distances was one of those inconvenient truths that had tended to be forgotten in the great Land Rover legend. However, there was no doubt that the forward-facing seats developed for 2007 were initially something of a shock for Land Rover traditionalists.

Essentially, the four inward-facing seats were replaced by two forward-facing seats that folded up against the sides of the vehicle when not in use. The arrangement was similar to that used in the 1998–2004 Discovery Series II models. Defender 90 Station Wagons came with these two seats as standard, while for Defender 110s they were made an extra-cost option. The front seats were changed for the 2007 models as well and gained taller backrests. On 110 Station Wagons, the rear bench was also redesigned, with a ⅓–⅔ split and a spring-assisted fold mechanism. As on the last of the Td5 models, there were three types of upholstery. The entry-level models still came with grey PVC, there was grey cloth for the County specification and its overseas equivalents, and the top level was called part-leather, featuring Ripstop cloth centres with leather-trimmed bolsters.

The 2007 Model Year

The 2007-model Defenders were announced at the Paris Motor Show in September 2006, although there had been a media launch in May that year, featuring pilot-production examples. Even so, the vehicles would not go on sale until February 2007, in line with the original target date for the project. The first production model (as distinct from pre-production samples) was actually built on 12 February 2007 and was on chassis number SALLDHAS77A-732644.

There were some new paint colours for 2007, plus a new model called the 110 Utility Station Wagon. There was nothing radically new about this, as all the components had been in place to build such a vehicle before. In practice, it was a 110 Station Wagon with plain 'van' sides in place of the rearmost side windows. With five seats in the front and a secure, covered cargo area in the back, Land Rover believed it would give them an alternative to the 110 Double Cab in the commercial market. Competition here had recently hotted up as Japanese 4×4s like the Nissan L200 and Toyota Hi-Lux had begun to offer stylish add-on van backs for their pick-up beds, so the new model was available with base, County and XS trim levels. Somewhat confusingly, it was marketed in sales catalogues as a Utility Pack option on the 110 Station Wagon.

LEFT: **First one! There was a ceremony for Solihull staff on the Defender lines. Flanking the 2007 model are HUE 166, the first pre-production model from 1948, and one of three identical V8 110s specially prepared for the *Lara Croft: Tomb Raider* film.**

BELOW: **The bonnet bulge is clear on this 2007 model, which has the XS specification with Brunel grey headlamp surround panels and the latest alloy wheel option. The bulging 'snout' reveals that it also has air-conditioning.**

RIGHT: **The angled flash underneath the Defender decal at the rear came in with the 2007 models.**

The forward-facing rear seats could be folded up against the body sides to give more room for cargo.

STANDARD COLOURS AND TRIMS

2007 and 2008 Model Years

Paints

Cairns Blue	LRC849
Rimini Red	LRC889
Chawton White	LRC603
Stornoway Grey	LRC907
Java Black	LRC697
Tonga Green	LRC904
Keswick Green	LRC799
Zermatt Silver	LRC798

On base models, wheel-arch eyebrows were in black (unpainted) and roofs were painted in white. On County and XS models, wheel-arch eyebrows and roofs were painted to match the body. However, all 110 hardtops, regardless of body colour, had a white roof; for these purposes, the 110 Utility Station Wagon was treated as a Station Wagon rather than a hardtop model.

Upholstery and Trim

Dark Granite Twill pattern PVC upholstery was used on base models, with grey Stag Vinyl for seat backs and sides. County derivatives had Black Span cloth upholstery with Black Mondus cloth for seat backs and sides. The 'part-leather' option had Black Rack 3 cloth with Black leather for seat bolsters, sides and backs.

The 2008 Model Year

The 2008 model year was dominated by publicity associated with Land Rover's celebrations of its own sixtieth anniversary. The Puma-engined models had only entered production in February 2007 and there was no reason to change their specification for the 2008 model year, which began just seven months later in September. However, the 2008 model year did see the first appearance of several new features that would gradually find a more permanent place in the range and most of these were featured on a special edition called the SVX. That name had been used before, on the Td5-engine concept vehicle built for the 1999 Frankfurt Motor Show, but Land Rover thought it was good enough to use again on a production model.

In fact, 2008 was to see a major publicity assault from Land Rover. To back up the sixtieth anniversary events, the company planned to provide major support for the Association of Red Cross and Red Crescent Societies worldwide. Land Rovers had been used as ambulances all round the world almost since the very beginning of the marque and the link was expected to deliver additional publicity for Land Rover as well as to support the work of the Association.

On 9 July 2008, Land Rover donated sixty brand-new vehicles to the Red Cross for use in the UK and overseas, arranging the actual handover to coincide with a Royal Garden Party at Buckingham Palace. The official presentation was made to the President of the Red Cross, HRH The Prince of Wales. The vehicles, mostly Defenders but also including Freelander 2 and Discovery 3 models, were arranged in the form of a Red Cross flag in the palace quadrangle. As for the garden party, it was one of the wettest for a long time.

The 2009 Model Year

It was fortunate that Land Rover had not planned any major product actions for the Defender for the 2009 model year.

LEFT: **Defenders have always been popular vehicles with relief agencies – this 2008-model 110 Station Wagon was pictured on duty with a Red Cross emergency response team during the Cockermouth floods in November 2009.**

BELOW: **Defenders figured prominently among the sixty vehicles that Land Rover donated to the Red Cross in 2008. The handover ceremony was held at Buckingham Palace.**

There had been signs that all was not well with the global economy by the end of 2007, but in September 2008 things took a sharp downward turn. Late 2008 and early 2009 were to have seen the third Land Rover G4 Challenge, a hugely expensive global sports adventure in connection with which Land Rover promised to donate £1 million to Red Cross funds. However, the financial prognosis for the immediate future was so bleak that Land Rover had no alternative but to cancel the event in December 2008. As a result, a number of specially prepared Defenders in the Tangiers Orange associated with the G4 Challenge appeared on the market in the first half of 2009. They would have been support vehicles for the athletes and buses for the invited media if the event had gone ahead as planned.

Some changes in UK commercial vehicle classification resulted in the 110 Utility Station Wagon being reclassified from an M1 passenger-carrying vehicle to an N1 light commercial with effect from 1 April 2009. The main effect of this

This Defender 130 with LRSV-built Quadtec body was built for the cancelled 2008–2009 G4 Challenge event.

This XS Station Wagon was pictured at the Canary Wharf International Expo in summer 2009.

was to allow those who used such vehicles as light good carriers to reclaim VAT on the purchase price.

Nevertheless, Land Rover did have plans for new Defender variants in hand and in the summer of 2009 it announced a very colourful limited edition that it hoped might help sales. This was the Defender Fire and Ice, which picked up on the vogue for two-tone colour schemes that had been pioneered by the BMW Mini, among other models. However, Land Rover's marketing people were convinced – probably rightly – that it was too garish for UK tastes and it did not appear in its home country.

The 2010 Model Year

By this stage, the Defender had more or less settled into a two-year model upgrade cycle, so 2010 was a fallow year after 2009. The promotional and marketing focus was on other models in the Land Rover range. Nevertheless, there was interest through limited editions in France, Germany and the Netherlands.

There was interest in Australia, too, where plans were afoot to try selling Defender 90s again. Behind the scenes, Land Rover Australia had been attempting to negotiate a better

STANDARD COLOURS AND TRIMS

2009 and 2010 Model Years

Paints

Alaska White	LRC603
Santorini Black	LRC820
Galway Green	LRC821
Stornoway Grey	LRC907
Izmir Blue	LRC920
Tamar Blue	LRC817
Keswick Green	LRC799
Zermatt Silver	LRC798
Rimini Red	LRC889

On base models, wheel-arch eyebrows were in black (unpainted) and roofs were painted in white. On County and XS models, wheel-arch eyebrows and roofs were painted to match the body. However, all 110 hardtops, regardless of body colour, had a white roof; for these purposes, the 110 Utility Station Wagon was treated as a Station Wagon rather than a hardtop model.

Upholstery and Trim

Dark Granite Twill pattern PVC upholstery was used on base models, with grey Stag Vinyl for seat backs and sides. County derivatives had Black Span cloth upholstery with Black Mondus cloth for seat backs and sides. The 'part-leather' option had Black Rack 3 cloth with Black leather for seat bolsters, sides and backs.

STANDARD COLOURS AND TRIMS

2011 Model Year

Paints

Baltic Blue	LRC912	From February 2011
Fuji White	LRC867	
Galway Green	LRC821	
Keswick Green	LRC799	
Rimini Red	LRC889	
Santorini Black	LRC820	
Stornoway Grey	LRC907	
Tamar Blue	LRC817	
Zermatt Silver	LRC798	

On base models, wheel-arch eyebrows were in black (unpainted) and roofs were painted in white. On County and XS models, wheel-arch eyebrows and roofs were painted to match the body. However, all 110 hardtops, regardless of body colour, had a white roof; for these purposes, the 110 Utility Station Wagon was treated as a Station Wagon rather than a hardtop model.

Upholstery and Trim

Dark Granite Twill pattern PVC upholstery was used on base models, with grey Stag Vinyl for seat backs and sides. County derivatives had Black Span cloth upholstery with Black Mondus cloth for seat backs and sides. The 'part-leather' option had Black Rack 3 cloth with Black leather for seat bolsters, sides and backs.

wholesale price for its dealers so that the Defender 90 could be priced more aggressively in the showrooms.

The news was announced in November 2009; the 90s, in Station Wagon form with electric windows and a CD player as standard, duly went on sale in early 2010. Land Rover gained maximum publicity with its target buyers through its sponsorship of the Quicksilver Pro surfing championships and arranging for the Quicksilver team to drive 90s from their headquarters in Torquay, Victoria, to the contest in Coolangatta on the Gold Coast over a period of several weeks. Early sales volumes looked promising.

The 2011 Model Year

The main focus at Land Rover in autumn 2010 was the pre-production publicity of the new Range Rover Evoque (which would not go on sale until autumn 2011), so changes to the Defender were consequently minimal. The major one was a realignment of the County and XS specifications. County models now came with the Brunel silver grille and headlamp surrounds first introduced on the XS, but some subtle rearrangements of the rest of the specification created a larger gulf between County and XS.

To the County specification, the XS now added the ABS-plus-ETC package, metallic paint and side runners. Air-conditioning was standard, together with heated front seats and windscreen, a stowage net on the tail door and a single-slot CD player with MP3 compatibility. The upholstery was in Black Rack 3 cloth with leather bolsters; there was a cubby box between the two front seats; and the steering wheel and gear-shift grips were covered in leather. The XS specification for the Defender 110 included the third row of seats, which made it a seven-seater.

Some options had now been moved to the new 'Professional Packs', consisting of an Exterior Pack, a Comfort Pack and a Style Pack. A new option for 90 and 110 derivatives was the Dual Finish alloy wheel style that had been previewed on limited editions during 2010. But 2011 brought its own new wheels as well, once again available only on limited editions. The new wheel was a gloss-black saw-tooth design, which helped to characterize the Defender Rough (for Europe but not the UK) and the latest X-Tech commercial special edition (for the UK). Both of these made their appearance in January 2011.

ABOVE: **This 110 XS Station Wagon belonged to the Land Rover Experience fleet that was used for driving tuition and demonstration work. The winch and checker-plate panels were not standard wear.**

RIGHT: **A heavily accessorized 2011-model Defender 90 Station Wagon shows off that year's new alloy wheels.**

The Next Stage – The 2.2-Litre Engine

There had been rumours for many years about the Defender's demise and Land Rover had more than once investigated the development of a replacement model. But nothing had gone beyond the early prototype stage, because in the end there had always been more pressing demands on the engineering budget.

However, by the start of the twenty-first century's second decade, it was clear that the Defender then in production could not continue much longer. New legislation was making the vehicle increasingly difficult to sustain and, icon or not, it was old-fashioned in comparison with its rivals. In September 2011, Land Rover unveiled a drivable concept model of what it expected to be the new Defender and suggested it would enter production sometime around 2015.

Whether the existing model could be kept in production for markets that did not apply restrictive legislation was something on which no company spokesman would be drawn. Nobody actually knew and what happened would probably depend on the situation at the time. However, a month before the announcement of the new Defender concept (code-named DC100 – Defender Concept, 100in (2,540mm) wheelbase), the company had modified the existing Defender to meet the challenges of the next few years.

The first challenge, and the primary reason for the model change, was the new EU5 diesel-emissions regulations that were due to come into force in 2011, after being postponed from their proposed 2009 date. The existing 2.4-litre Puma engine could not be modified to meet the regulations and still deliver acceptable performance, but the Land Rover engineers pulled a surprising solution out of the bag. In place of the 2.4-litre Puma engine, they fitted its small-bore 2.2-litre relative, an engine introduced at the same time in the 2012 Ford Transit. Confusingly, there was also a 2.2-litre Ford turbocharged diesel in the Freelander 2 and Range Rover Evoque models, but that was not the same engine, being a transverse design developed in tandem with Peugeot in France.

Perhaps the most impressive feature of the new engine was that it delivered the same power and torque as the old one despite having a smaller capacity and a Diesel Particulates Filter (which could be omitted on Defenders for non-EU5 markets). Land Rover's own publicity put this remarkable achievement down to an improved combustion system and a sophisticated engine-management system. Fuel economy remained the same as with the 2.4-litre engine, but users had extra benefits. First, the new engine was actually quieter than the old, thanks to a full acoustic cover instead of the 2.4's simple splash cover. Second, the engine-management system now allowed a more respectable top speed of 90mph (145km/h).

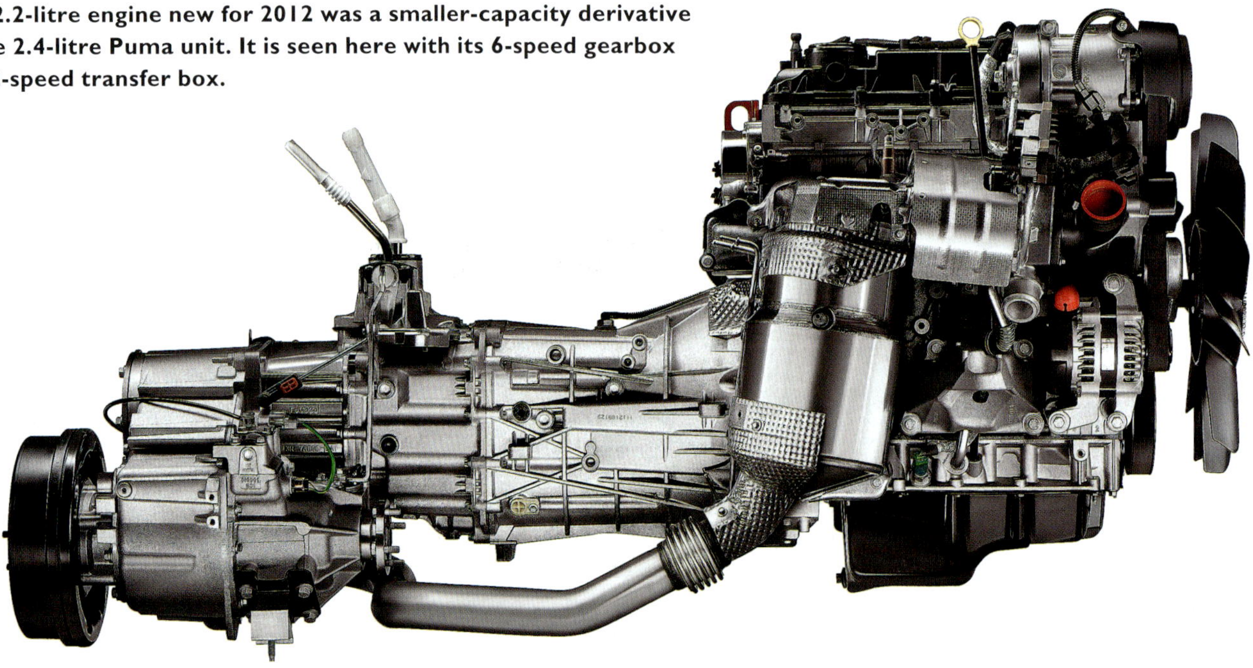

The 2.2-litre engine new for 2012 was a smaller-capacity derivative of the 2.4-litre Puma unit. It is seen here with its 6-speed gearbox and 2-speed transfer box.

The 2012 Model Year

The 2012 model-year Defenders were announced on 12 August 2011, an early date chosen so that the introduction of the DC100 concept vehicle – promoted as anticipating the 2015 Defender replacement – would have a clear run at publicity in the media when it was introduced at the Frankfurt Show in September.

With the new 2.2-litre engine came some changes to the three speed-limiting options that were favoured by fleet buyers as a way of containing fuel costs. The 2012 models had the same MT82 six-speed gearbox as before, albeit with a lower first gear, and came with the same range of models in the same three-tier structure of base, County and XS. They also saw the standardization of ventilated front disc brakes on Defender 90 models, so bringing their braking systems into line with those on 110s and 130s.

There were some changes to the paint options, of course, and further changes here followed at the start of the 2012 calendar year. Tinted glass was now a standard feature on all models. Customer feedback had led Land Rover to offer some alternative hood colours as well, and black became an option for the pick-up models while beige became an option for the 110 Double Cab. The gloss black Saw Tooth types seen on earlier special edition models were now part of the main options list, along with the Dual Finish five-spoke alloy wheels introduced for 2011.

There was also some shuffling of option packs, and for 2012 two new ones were available. The Comfort Pack brought electric windows, remote central locking, air conditioning, and a CD player with auxiliary input. The Off-Road Pack brought heavy-duty wheels with Goodyear MT/R tyres, a tow ball, a front undershield and ABS with ETC.

ABOVE: **Saw-tooth alloy wheels with a gloss black finish were added to the 2012 options list.**

RIGHT: **This 2012-model 110 Double Cab has the latest black tilt cover, an XS trim level with air-conditioning and Dual Finish alloy wheels, plus checker-plate panels and a winch. The Dual Finish wheels were not advertised as available for 130 models.**

The XS trim level and air-conditioning are again apparent on this 2012-model 90 Station Wagon.

STANDARD COLOURS AND TRIMS

2012 Model Year

Paints

Aintree Green	LRC866	From early 2012
Firenze Red	LRC868	
Fuji White	LRC867	
Galway Green	LRC821	To early 2012 only
Indus Silver	LRC863	
Keswick Green	LRC799	
Orkney Grey	LRC949	
Santorini Black	LRC820	To early 2012 only
Sumatra Black	LRC797	From early 2012
Tamar Blue	LRC817	

On base models, wheel-arch eyebrows were in black (unpainted) and roofs were painted in white. On County and XS models, wheel-arch eyebrows and roofs were painted to match the body. However, all 110 hardtops, regardless of body colour, had a white roof; for these purposes, the 110 Utility Station Wagon was treated as a Station Wagon rather than a hardtop model.

Upholstery and Trim

Dark Granite Twill pattern PVC upholstery was used on base models, with grey Stag Vinyl for seat backs and sides. County derivatives had Black Span cloth upholstery with Black Modus cloth for seat backs and sides. The 'part-leather' option had Black Rack 3 cloth with Black leather for seat bolsters, sides and backs

The 2013 Model Year

The 2013 model-year was always going to be overshadowed at Solihull by the introduction of the fourth-generation Range Rover, coded L405. Therefore, no major changes were planned for the Defender as it entered its thirtieth year in production. Nevertheless, the model-year kicked off early with a brightly coloured limited edition in July 2012 (see Chapter 10).

The real 2013 models were nevertheless not available until September, their public announcement being on 30 August at the Burghley Horse Trials that Land Rover sponsored. Earlier special edition models had experimented with roofs in contrasting colours other than white, and now four new contrast colours became available. This move towards two-toning of course reflected the stylish two-toning available since autumn 2011 on the Range Rover Evoque. There were two new metallic body colours available, too: Barolo Black and Havana.

Also pioneered on the special editions had been Recaro bucket seats at the front, and these were now added to the options list. There were ICE upgrades, too: the base specification consisted of a single-slot CD player with USB plug and auxiliary jack, and four 50-watt speakers; an optional upgrade for this added Bluetooth connectivity with a remote microphone; and a new Hi Line Alpine 150-watt system was available with a low-profile sub-woofer and 10cm co-axial speakers in place of the standard types.

However, the highlight of the 2013 model-year would be the production of the two millionth model – billed as the two millionth Defender although it was the two millionth workhorse Land Rover model since 1948, and not the two millionth to carry the Defender name. As this book went to press, the event was scheduled for March 2013, and an associated limited edition was on the cards for later in the year.

For the 2013 models, alternative contrasting roof colours became available. Front seats were also changed, and the ICE options were upgraded.

STANDARD COLOURS AND TRIMS

2013 Model Year

Paints

Aintree Green	RC866	
Barolo Black	LRC861	Additional 'Premium' colour
Firenze Red	LRC868	
Fuji White	LRC867	
Havana	LRC865	Additional 'Premium' colour
Indus Silver	LRC863	
Keswick Green	LRC799	
Orkney Grey	LRC949	
Sumatra Black	LRC797	
Tamar Blue	LRC817	

On base models, wheel arch eyebrows were in black (unpainted), but on County and XS models they were painted to match the body. Roofs could be painted to match the body or in a contrasting colour: the roof colours available were Firenze Red, Fuji White, Indus Silver, Orkney Grey and Santorini Black.

Upholstery and trim

Dark Granite Twill pattern PVC upholstery was used on base models, with grey Stag Vinyl for seat backs and sides. County derivatives had Black Span cloth upholstery with Black Modus cloth for seat backs and sides. The 'part-leather' option had Black Rack 3 cloth with Black leather for seat bolsters, sides and backs.

2007–13 Models

(i) European-Type VINs

The VIN codes used in this period consisted of 17 characters, made up of an 11-character prefix code and a six-digit serial number.

Example: SALLDHAS77A-xxxxxx

This breaks down as follows:

SAL Manufacturer code (Land Rover)

LD Defender

H 110in (2,794mm) wheelbase
- K = 127in (3,226mm) wheelbase (Defender 130)
- V = 92.9in (2,360mm) wheelbase (Defender 90)
- W = Defender 90 heavy-duty (2011 MY onwards)

A Truck cab, soft-top or hardtop body (to end 2010 MY)
- B = 90 Station Wagon
- F = Crew Cab with standard pick-up body (i.e. 110 Double Cab)
- H = HCPU body with truck cab or crew cab (to end 2010 MY)
- L = Soft-top (2011 MY onwards)
- M = 110 Station Wagon
- N = Hardtop (2011 MY onwards)
- R = Truck cab (2011 MY onwards)
- S = Crew cab (2011 MY onwards)
- T = HCPU body with truck cab (2011 MY onwards)
- V = HCPU body with crew cab (2011 MY onwards)
- W = Chassis-cab with truck cab (2011 MY onwards)
- X = Chassis-cab with crew cab (2011 MY onwards)
- Y = 110 Utility Station Wagon (2011 MY onwards)

S 2.4-litre TDCi (Puma) engine to EU4 specification
- N = 2.4-litre TDCi (Puma) 24V military engine to EU4 Specification
- P = 2.2-litre TDCi (Puma) engine to EU5 specification
- R = 2.2-litre TDCi (Puma) engine to Rest of the World Specification
- T = 2.4-litre TDCi (Puma) engine to Rest of the World Specification
- W = 2.4-litre TDCi (Puma) 24V military engine to EU2 Specification for the Rest of the World

7 RHD with 6-speed manual gearbox
- 8 = LHD with 6-speed manual gearbox

7 Model year 2007

8 = 2008	A = 2010	C = 2012
9 = 2009	B = 2011	D = 2013

A Assembled at Solihull
- C = Assembled in Zimbabwe
- D = Assembled in Thailand
- F = Shipped as KD for unidentified assembly location
- J = Assembled in Malaysia
- K = Assembled in Kenya
- N = Assembled in Morocco
- P = Assembled in Pakistan
- R = Assembled in Jordan
- V = Assembled in South Africa
- W = Assembled in Turkey

xxxxxx Serial number

(ii) NAS-Type VINs (Used in Japan and Certain Other Territories)

The VIN codes used in this period consisted of seventeen characters, made up of an eleven-character prefix code and a six-digit serial number.

Example: SALDVBS717A-xxxxxx

This breaks down as follows:

SAL Manufacturer code (Land Rover)

D Defender

V 92.7in (2,355mm) wheelbase (i.e. Defender 90)
- H = 110in (2,794mm) wheelbase

VINS 2007–13 Models

B	Three-door body (90 Station Wagon) M = Five-door body (110 Station Wagon)
S	2.4-litre TDCi (Puma) engine to EU4 specification
7	RHD with 6-speed manual gearbox 8 = LHD with 6-speed manual gearbox
1	Security check digit (0 to 9, or X)
7	Model year 2007 8 = 2008 A = 2010 C = 2012 9 = 2009 B = 2011 D = 2013
A	Assembled at Solihull
xxxxxx	Serial number

Engine Identification Numbers for Defender, 2007–2013 models

The Puma engines have Ford identification numbers, which are completely different from the earlier Land Rover system. The numbers include data to represent build date and time as well as specification. Note that the prefix ELD indicates an engine built at Dagenham.

As an example, the first production engine number for 2007-model Defenders was ELD0006111911063l.

SPECIFICATIONS DEFENDER 2007–13 on

Diesel Engines

Type	Puma 2.4-litre
Bore and stroke	89.9mm × 94.6mm
Capacity	2,401cc
Valves	4-cylinder with 4 valves per cylinder
Compression ratio	17.5:1
Carburettor	common-rail injection by solenoid-driven injectors, and Garrett variable-geometry turbocharger
Max. power (DIN)	122bhp at 3,500rpm and 265lb ft at 2,000rpm

Type	Puma 2.2-litre
Bore and stroke	86mm × 94.6mm
Capacity	2,198cc
Valves	4-cylinder with 4 valves per cylinder,
Compression ratio	17.5:1
Carburettor	common-rail injection by solenoid-driven injectors and Garrett variable-geometry turbocharger
Max. power (DIN)	122bhp at 3,500rpm and 265lb ft at 2,000rpm

Transmission

Permanent four-wheel drive

Main gearbox

6-speed MT82 manual gearbox

Ratios (2.4-litre models) 5.443:1, 2.839:1, 1.721:1, 1.223:1, 1.000:1, 0.742:1, reverse 4.935:1; (2.2-litre models) 5.441:1, 2.839:1, 1.721:1, 1.223:1, 1.000:1, 0.742:1, reverse 4.89:1.

Transfer gearbox

2-speed

Ratios 1.211:1 (high) and 3.269:1 (low)

Axle ratio: 3.54:1

Suspension and Steering

Suspension

Coil springs all round

Steering

Worm and roller steering with 20.2:1 ratio; optional power assistance with 19.3:1 ratio; PAS standard

Continued overleaf

Continued from previous page

SPECIFICATIONS | DEFENDER 2007–13

Brakes

Type Disc brakes on all four wheels; some heavy-duty models with ventilated front discs; servo assistance; separate drum-type transmission parking brake

Electrical System

12V with alternator

Dimensions

90 models

Track	58.5in (1,486mm)
Wheelbase	92.9in (2,360mm)
Overall length	146.5in (3,720mm)
Overall width	70.5in (1,791mm)
Unladen height	77.3in (1,963mm) for truck cab 77.4in (1,966mm) for soft-top 77.6in (1,971mm) for hardtop and Station Wagon
Unladen weights	3,611lb (1,638kg) minimum with Puma 2.4 diesel engine 3,836 lb (1,740kg) minimum with Puma 2.2 diesel engine

110 models

Track	58.5in (1,486mm)
Wheelbase	110in (2,794mm)
Overall length	175in (4,445mm) for truck cab and soft-top 180.3in (4,580mm) for Station Wagon 184in (4,674mm) for HCPU
Overall width	70.5in (1,791mm)
Unladen height	80.1in (2,034mm)
Unladen weights	3,858lb (1,750kg) minimum with Puma 2.4 diesel engine 4,226lb (1,917kg) minimum with Puma 2.2 diesel engine

130 models

Track	58.5in (1,486mm)
Wheelbase	127in (3,226mm)
Overall length	1,98in (5,029mm)
Overall width	70.5in (1,791mm)
Unladen height	80.1in (2,034mm)
Unladen weights	4,425lb (2,007kg) minimum with Puma 2.4 diesel engine 4,738lb (2,149kg) minimum with Puma 2.2 diesel engine

Performance

90 with Puma 2.4-litre engine and 6-speed MT82

0–60mph	(No data available)
Maximum	82mph (132km/h)
Fuel consumption	(No data available)

90 with Puma 2.2-litre engine and 6-speed MT82

0–60mph:	(No data available)
Maximum:	90mph (145km/h) 25–28mpg (11.3–1ltr/100km)

CHAPTER TEN

THE VALUE OF SCARCITY

Special Editions and Limited Editions

There was a major change in Land Rover's approach to its workhorse models at the start of the 1990s. Ever since the introduction of the original Land Rover in 1948, the vehicle had sold itself and the issue of 'marketing' simply did not come into the equation. However, as the leisure 4×4 market grew, it became clear that there were potential sales among customers whom Land Rover had never previously considered. By the close of the 1990s, the job of identifying new groups of customers ('market sectors') and developing Land Rover derivatives to suit them was given to a team within the Marketing Department. This team was known as Defender Brand Management, after the new name for the workhorse models that arrived that year.

The major sales opportunity for Defender was quickly identified as the leisure market itself and the Brand Management team set about dreaming up new derivatives of existing models that could be produced to suit it. Cost was a major issue, of course, and another part of the Brand Management team's job was to ensure that no derivative would have a negative impact on sales of others (for example, by weakening the image of Defender as a rugged off-road workhorse). Another issue was how to get new 'leisure' derivatives manufactured if they had marked differences from the regular line-production models. That issue was solved when Land Rover Special Vehicles (*see* Chapter 1) came on-stream during 1992. Among the reorganized division's first projects was the first limited-edition Defender for the UK, the 90SV.

Outside the UK, some of Land Rover's National Sales Companies could not wait for more special-edition models to be produced. They began to create their own, usually by specifying an attractive combination of special features from the options list and sometimes by adding special decals, badges or other unique but inexpensive items. Leaders in this field were Land Rover France, who developed a series of very successful special-edition and limited-edition models during the 1990s and into the 2000s as well. Other European countries followed suit and during the 2000s there were both 'global' limited editions designed to suit several countries and country-specific limited editions, all made at Solihull. There were even some special editions where different derivatives of the same edition were sold in different countries.

The 1992 Model Year

Defender 90 Trophy Special Edition (France)

To Land Rover France went the honour of creating the very first special-edition Defender, which went on sale in April 1992. The base vehicle was a white Defender 90 Tdi with the French 'hardtop' specification, which was essentially a Station Wagon with no rear seats to avoid taxation.

The Defender Trophy had a body-colour grille and wheel-arch eyebrows (as on UK County models), white enamelled Rostyle wheels and the standard white roof. This all-over white colour scheme was complemented by a black nudge bar carrying auxiliary driving lamps, black light grilles front and rear and black side runners. A sunroof and tow-hitch were standard and there were three front seats upholstered in Moorland cloth.

The unique features were a yellow and grey side stripe along the body shoulder feature, with yellow door decals reading 'Defender Trophy' and looking very much like the Camel Trophy logos that had clearly inspired them. It is not

With its full rear roll cage, the SV90 never looked as exciting as the NAS 90, even though they shared a common ancestry. The 200Tdi diesel engine took the edge off the excitement potential, too, but the colour was certainly eye-catching. Both these pictures show one of the prototypes; it had neither rear seats nor production-style padding on the roll cage, and there had been paint-matching problems on the panels as well.

clear how many of these vehicles were built, but the Defender Trophy was still available through French dealers in July 1993, by which date the price had increased considerably.

The 1993 Model Year

Defender 90SV (UK)

The 90SV was a critical element in changing public perceptions of the Defender in the UK. The Defender Brand Management team led by Mike Gould were aiming for a low-volume special-edition model with a high profile to kick-start the process, and in the 90SV they got it. What they ended up with, however, was not the vehicle they had started work on. Both their concept prototypes were based on 90 hardtops, one being a deliberately Spartan, stripped-out model and the other more luxurious and better-equipped. The latter evolved into the 90 County hardtop, launched in autumn 1992.

Contact with LRSV, who would have to build the vehicle, led them to the project team working on the vehicle that would eventually become the NAS 90 (see 1994 model year below). This brought about a rethink. It made sense to share project costs and the UK plan evolved to focus on a summer-oriented soft-top with large helpings of style and glamour. The team wanted the same V8 petrol engine as the NAS 90, but no catalyst-equipped exhaust had been developed for RHD V8 Defenders and the cost of engineering a special one for a planned run of just ninety vehicles ruled it out. So the project went ahead with the much less glamorous 200Tdi diesel, which failed to deliver the performance that the model's appearance suggested.

From the NAS 90, the project inherited the padded rollover bar and its outside section around the windscreen. It also inherited the ideas of replacing the bulkhead behind the front seats with a bracing bar and of using five-spoke alloy wheels, although the plan to fit chunkier tyres did not receive Engineering approval in time for the UK vehicle. Acknowledging the involvement of Special Vehicles, that the vehicle was based on a Defender 90 and that there were to be ninety of them, the limited edition took on the name of 90SV.

It also took on a bright, summery blue-green metallic paint called Caprice, which would later appear on Discoveries but remained unique on Defenders. A front A-frame protection bar, twin-tube side runners and a swing-away spare wheel carrier at the rear (borrowed from an unrelated military programme) nearly completed the picture. The last thing was a removable soft-top – needed because summer rarely lasts long in the UK – and for that Land Rover went to Tickford of Newport Pagnell. Tickford's design for a sophisticated double-skinned cotton soft-top was superb, but it relied upon the base 90s being built to closer tolerances than Land Rover could achieve in production. So a redesign introduced delays and, although the 90SV was announced in July 1993 at the height of summer, volume production did not begin until August. By then, the 90SV had missed the summer market. That and a high price deterred buyers, and around half to the eighty-eight built (plus two prototypes) were shipped to Tanzania and South Africa. Even so, the 90SV beat the related NAS 90 into the showrooms because delays on that project held it up.

Special features of the 90SV included a colourful, summery decal badge on each front wing and on the tailgate; one model was finished with larger decals based on this design for the September 1993 London Motor Show. All 90SVs were built under LRSV contract number 5834 and this is recorded on a plate under the bonnet, which also gives the number of each vehicle within the production run and the date of its completion by LRSV. In practice, the 90SVs were built in small batches and there was some evolution of details between one batch and the next.

Optional extras for these models were twin driving lights, a winch with a special front bumper, a tow pack and a four-speaker RDS stereo radio-cassette ICE system with 6-CD autochanger in place of the standard radio-cassette unit with two speakers.

The 1995 Model Year

Defender 110 Tanami Limited Edition (Australia)

Land Rover Australia created a special edition to promote the new 300Tdi diesel engine, marketing it in association with Les Hiddins. Les Hiddins was well known as the Bush Tucker Man, with a TV programme about survival in the outback that focused on finding 'bush tucker' – food.

There were no short-wheelbase models available in Australia at the time and the special edition was based on a Defender 110 Station Wagon. It was called the Tanami, after the desert in northern Australia. The base vehicle was painted white and carried full-length 'desertscape' decals on its flanks with the Tanami name on each front door. At the back of

The Bush Tucker Man himself, Les Hiddins, posed for the publicity pictures of the Defender Tanami.

each rear wing was a decal of Les Hiddins' unmistakable bush hat, together with a decal of his signature. Special equipment consisted of a black bull bar and snorkel, a full-length roof rack with rear ladder and a pioneer tools set of shovel, axe and pick mounted on the bonnet.

The total made was just thirty. Records show that two were painted red and one was painted green; all the others were white.

The 1996 Model Year

Defender 90 Eastnor Special Edition (France)

Land Rover France created a special edition with a summer theme that it called the Eastnor, after Land Rover's famous testing ground in Herefordshire, and announced it in February 1996. The model was similar in most respects to the UK-market 90SV of nearly three years earlier, with some elements of the NAS 90 added into the mix and a few unique elements as well. There were 100 examples, all based on Tdi-engined Defender 90 soft-tops with full-height doors, a side-hinged tailgate and a four-point padded rollover bar.

The Eastnor was painted yellow, with black wheel-arch eyebrows and carried large 'Eastnor' decal logos on the bonnet and doors. It had a NAS-style rear bumper and Freestyle Choice wheels with 275/75 × 16 tyres. Standard equipment included a front A-bar carrying two long-range driving lamps protected by individual stone guards, black side runners and a rear step. There were mudflaps front and rear, and silver-finish chequer-plate on the wing tops. The interior featured three seats, all upholstered in Moorland cloth, with head restraints for the outer pair.

Most interesting was that the Eastnor also had a NAS-style demountable three-piece hardtop in Ash Grey GRP. It seems to have been the only non-NAS model to have this feature.

Defender 90 X-Trem Special Edition (France and Belgium)

One of the drawbacks of the Defender's soft-top configuration was the length of time needed to remove or replace the canvas top. In France, Land Rover specialists Challenge Concept, based in Dardilly, developed a simple folding top that could be bolted onto an open Defender 90 to turn it into a convertible. This was ideal for climatic conditions in the Mediterranean region of France and during 1996 Land Rover France took a quantity of these convertible tops for a special-edition model that they called the X-Trem (which translates directly as 'extreme').

There were 200 examples of the X-Trem for France, all based on Defender 90 Tdi soft-tops with the 300Tdi engine

The X-Trem was the first-ever Defender cabriolet and was ideally suited to the Mediterranean climate. All models carried Special Vehicles decals. The demountable hardtop above was a rare extra-cost option.

and all intended for sale in France. Belgium took a further 150 examples. They were available in Rioja Red or Riviera Blue, with black convertible tops. These were well-made items with a three-layer construction of rubber sandwiched between cotton. The X-Trem had large distinguishing decals on its doors, Freestyle Choice five-spoke alloy wheels with 275/75R16 tyres, side runners and a rear step, plus an A-bar with two long-range driving lamps protected by individual stone guards.

An optional extra available for the X-Trem was a polyester resin hardtop, made by Challenge Concept to its own distinctive design with an angled rear panel and two windows in each side separated by a pillar angled to match the rear panel. This normally had a black finish, but could also be painted to match the body at extra cost; a heated rear window was an extra-cost option and it was possible to order the hardtop without the side windows. This hardtop was available either through Land Rover dealers in France and Belgium or direct from Challenge Concept, but was quite expensive and remained rare.

The 1997 Model Year

Defender 90 Highlander Special Edition (France)

Land Rover France continued its aggressive programme of special-edition Defender 90s in October 1996 with the Defender 90 Highlander. No details of quantities are available.

The Highlander was a 90 Station Wagon in Epsom Green with the Freestyle Choice wheel, tyre and anti-roll bar option. The roof and wheel-arch eyebrows were both painted in the body colour (the roof being the main novelty at this stage), and the Highlander came with tinted glass all round and a removable sunroof. There were special Highlander logos as well. An alarm was standard and the radio head unit came with a removable faceplate.

The 1998 Model Year

Defender 90 Ultimate Limited Edition (France)

This limited edition from Land Rover France was not intended as part of the fiftieth anniversary promotional push, but as a separate attempt to exploit a particular market niche. On sale from October 1997, the Ultimate was originally intended as a fifty-strong limited edition, but strong demand led to the production of a further batch of eighty.

The specification was broadly that of a 90 truck cab with County features. The Ultimate came in Charleston Green metallic (which was not otherwise available on Defenders), with body-coloured roof, wheel-arch eyebrows and mirror bodies. The front wings carried 'Ultimate' logos and there was a four-point rollover bar behind the cab, together with a black tonneau cover for the pick-up bed. An A-bar at the front carried two long-range lamps with individual stone guards and a tow hitch was fitted. The Freestyle Choice alloy wheel, 275/75 Goodrich tyre and anti-roll bar package was standard, with four mudflaps. The cab had three seats upholstered in the latest grey Rayleigh cloth and there was a radio-cassette unit with removable faceplate and two speakers.

Defender 90 V8i 50 Collector's Edition (UK)

The keynote limited edition associated with Land Rover's fiftieth anniversary year was the Defender 90 V8i 50 which, as its name suggests, was powered by the latest injected 4.0-litre V8 petrol engine. It was in many respects a RHD derivative of the 1997 NAS 90 'hardtop' model and shared that model's rear-mounted petrol tank, but all 385 examples for the UK were painted in the special Atlantis Blue micatallic paint that was reserved for fiftieth anniversary editions of the Range Rover, Discovery and Freelander as well. The first examples were built in May 1998 and went on sale in June; marketing material described them as a Collector's Edition.

The V8 50 shared its ZF four-speed automatic primary gearbox with the 1997 NAS 90 and was in the same state of tune. The Station Wagon body had a body-coloured roof and wheel-arch eyebrows, with an external roll cage and a raised V8 badge on each front wing. There was an aluminium limited-edition plate on the tail panel, with a green and gold oval bearing the fiftieth anniversary logo. The plate identified the vehicle by its number within the limited edition and preceded this by the letters GB (for Great Britain). Aluminium-finish chequer-plate panels were used on the wing tops, the sills, the body's lower rear corners and the rear step, and there were stainless-steel side runners.

Particularly effective was the use of a two-tone version of the five-spoke wheels associated with the Freestyle Choice package, with gunmetal spokes contrasted against bright silver rims. The full Freestyle Choice package of chunky BF Goodrich 265/75R16 tyres and anti-roll bars was also fitted and it seems to have been about this time that the five-spoke wheels were first described by the Tornado name.

Inside the cabin, there were leather upholstery, a leather-bound steering-wheel rim and a leather-bound grip for the automatic transmission selector. The grips for the window

The V8 50 edition for the UK was a real show-stopper, with special Atlantis Blue paint, two-tone wheels and of course the V8 engine and automatic gearbox. Each one carried an individually numbered plaque.

winders and transfer-box lever were meanwhile made of cast aluminium. The centre console carried a miniature version of the limited-edition plate, with the vehicle's identifying number. Air-conditioning was standard and the stereo radio-cassette unit was accompanied by four speakers. A six-CD autochanger was available (for the first time in the UK) as a dealer-fit accessory.

Defender 90 V8 50 Limited Edition (Outside the UK)

There were derivatives of the V8 50 limited edition for some overseas markets as well, but the precise specification varied to suit local market tastes. All of them carried a numbered limited-edition plaque on the right-hand tail panel.

The most numerous of the special fiftieth anniversary editions was the one for Japan. Like the others, it had a 4.0-litre V8 engine and automatic gearbox.

The most numerous derivative was built for Japan, which took 451 examples. These had Alpine White paintwork (a Japanese favourite), satin-finish Boost alloy wheels (as used on the Discovery in the UK) and grey cloth upholstery.

There were also 180 for Germany, twenty-five for France, twenty for the Netherlands and ten for Belgium. The overall total of V8 50 models worldwide, including the UK models, was therefore 1,071.

Defender 90 Tdi 50 Hardtop Limited Edition (UK)

The name of this limited edition was hardly memorable or inventive, but it capitalized on the promotional activities of Land Rover's fiftieth anniversary to sell a highly specified Defender 90 hardtop. It was the first UK limited edition based on a commercial rather than a passenger-carrying model and all 200 were built in July 1998 for an August start to sales.

All models had the metallic paint associated with the County specification, with the roof and wheel-arch eyebrows painted in the body colour. There were eighty examples in Cobar Blue, seventy in Rioja Red and fifty in Woodcote Green, and there was a chequer-plate body protection kit. No side decals were used and there was a traditional-style aluminium Land Rover logo on the tail panel in place of the standard green and gold item. The Freestyle Choice wheel, tyre and anti-roll bar package was fitted, but the wheels were two-tone gunmetal Tornado alloys like those on the V8 50.

The special features of a black leather-bound steering wheel and aluminium grips for the window winders and the main and transfer gear levers accompanied a County interior package.

The 1999 Model Year

Defender 50th (France)

There were no V8 50 models for France, where taxation made a 4.0-litre petrol engine unsaleable, so Land Rover France developed its own commemorative special edition for the marque's fiftieth anniversary. It was called the Defender 50th, and was based on a Td5-powered 90 Station Wagon.

There were 300 examples and sales began in October 1998. The Defender 50th came in Sand or Atlantic Green (a light green that would also be used on the UK-market Defender Heritage models). Roofs and Rostyle steel wheels were in white, while the wheel-arch eyebrows were black. There were '50th' logos on the wings and tail panel. Tinted glass was standard, together with four mudflaps, a tow hitch, a black tubular bull bar and auxiliary lights with stone guards. A radio-cassette unit was standard and the ABS-plus-ETC package was an extra-cost option.

Defender Heritage Edition (UK)

The Defender Heritage Edition was announced in August 1999 to catch the tail end of the summer sales season, but became available in September 1999. Land Rover treated it as a 1999 Model Year derivative and told its dealers that these vehicles were intended to appeal to 'Active Lifestyle owners'.

There were both 90 and 110 versions of the Heritage Edition, the 110 becoming the first special edition for the UK based on the long-wheelbase model. Both were Station Wagons and both had a unique metal mesh-effect grille that was intended to recall the wire-mesh grille of early Land Rovers. It appears that a real wire-mesh grille had been in the original proposal, but that it was rejected as too expensive. There were Land Rover badges in aluminium and silver-painted exterior details to suggest the galvanizing on older models.

The Heritage models came in either Bronze Green (the traditional Land Rover colour that had been deleted in 1988), or Atlantic Green (chosen to resemble the light green used on 1948 Land Rovers). Of the 270 examples built, there were 189 on the short wheelbase and eighty-one on the long wheelbase. Land Rover painted seventy-six of the 90s in Atlantic Green and 113 in Bronze Green; on the 110s, there were sixty-six in Bronze Green and just fifteen in Atlantic Green. The 90s had five-spoke alloy wheels with the spokes finished in the body colour and there were body-colour wheels for the Atlantic Green 110s, although the Bronze Green examples had bright-finish Freestyle Choice wheels.

A unique feature of the interior was leather upholstery in mid-green. There was originally to have been a leather-bound steering wheel as well and this was mentioned in the 'teaser' sales brochure but was deleted before production began, again on cost grounds. However, an aluminium gear-shift grip was standard, together with ivory faces and chrome bezels for the instrument dials. None of these was authentically 'early Land Rover', but they contributed to the old-fashioned air that the vehicles were trying to conjure up. Air-conditioning was an extra-cost option, as was the ABS-plus-ETC package.

Colours for the Heritage Edition were intended to be 'Land Rover traditional', but that fake grille did nothing for the vehicles' credibility.

The 2000 Model Year

Defender 90 X-Trem 2 (France)

Land Rover France continued its successful series of limited-edition Defender 90s with the X-Trem 2, following on from the first X-Trem in 1996. This was another cabriolet, using the latest version of the conversion developed by Challenge 4×4 (formerly Challenge Concept) in Dardilly, and was announced in May 2000 for the summer season. There were eighty examples, available in Alveston Red, Caledonian Blue, Chawton White, Coniston Green, Epsom Green, Java Black, Oxford Blue and Rutland Red. Some of these colours were not among the standard Defender options; the metallic paints cost extra. The convertible tops were in black or beige vinyl and there were X-Trem 2 logos on the doors.

These vehicles also had carbon-look protection plates on the front wings and rear lower corners, plus a front undershield. There was a simple rubber mat in the rear load space and the spare wheel was on a swing-away carrier at the rear, attached to the chassis. There was an A-bar at the front, carrying a pair of Hella long-range driving lamps and the Freestyle Choice wheel, tyre and anti-roll bar package was standard.

France's X-Trem 2 was essentially the same as the original X-Trem, but now had the Td5 engine.

Defender 90 X-Tech (UK)

The X-Tech limited edition was put together to test the water for special editions among customers for commercial rather than passenger-carrying Defenders. It was announced in August 2000 and Land Rover described it as a 2000 Model Year vehicle; many of the fifty built were sold in what was strictly the 2001 Model Year, however. Land Rover described it to its dealers as a 'working vehicle but with all the accent on Fun!'

The X-Tech was a Defender 90 hardtop in Blenheim Silver with County-style matching paint treatment for the wheel-arch eyebrows and roof. The wheels were dished alloys on 265/75R16 tyres and the ABS-plus-ETC package was standard. All examples had tinted glass, Techno cloth upholstery with a cubby box between the seats, carpets and a radio-cassette unit. They also had a leather-bound steering wheel rim and a polished aluminium gear-shift grip.

The first X-Tech in 2000 explored the possibilities of a special edition for the commercial market.

The 2001 Model Year

Defender 90 X-Tech 2 (UK)

Following the success of the X-Tech limited edition for the commercial market in autumn 2000, Land Rover built a further fifty for sale from April 2001. This second X-Tech edition had some minor differences from the first.

Once again, the base vehicle was a 90 hardtop in Blenheim Silver with body-coloured wheel-arch eyebrows and roof. This time there was a sunroof as well as Alpine lights and the tail panel carried a brushed aluminium X-Tech badge plate. The wheels were sculpted alloys with a bright finish. The ABS-plus-ETC package was standard.

The cab had the latest two-seats-plus-cubby configuration, with a carpeted floor, leather-bound steering wheel, aluminium gear-shift grip and radio-cassette unit. There was a dealer-fit extra-cost option called the X-Tech Kit, which consisted of chequer-plate protection panels and a stainless-steel A-bar that carried two Rally 1000 driving lamps.

The X-Tech carried a special-edition badge on its tail panel.

Defender 90 Hawaii (France)

Land Rover France announced its special edition for summer 2001 at the Salon du Coupé et du Cabriolet exhibition in Paris towards the end of March. There were 130 examples, all for France and all Defender 90 soft-tops in Monte Carlo Blue with beige window soft-tops. They had 'Hawaii' logos on the door tops, just behind the mirrors.

The Hawaii models came with ABS and ETC as standard, plus a Freestyle Choice wheel, tyre and anti-roll bar package. Tinted glass, a tow hitch and a radio-cassette unit were part of the package.

Defender Tomb Raider

The Defender Tomb Raider special edition was announced in June 2001 and was available globally as a 90 Station Wagon or 110 Double Cab. There were 1000 examples altogether, all finished in Bonatti Grey and all carrying special Tomb Raider identification badges. Their appeal was supposedly to the urban would-be adventurer, which explained their association with the new *Lara Croft: Tomb Raider* film that had its UK premiere on 6 July. Land Rover told its UK dealers that the Tomb Raider models were 'aimed at individuals who want to stand out from the crowd' and that they would be 'appealing to customers who drive high image vehicles'. Nevertheless, conquest sales were an important part of the strategy, too: the Tomb Raider 'may be some customers' first experience of the Land Rover brand'.

Lara Croft: Tomb Raider was a popular video game that had been turned into a film. Land Rover made a successful bid to supply vehicles for the heroine (played by Angelina Jolie) to drive in the film and LRSV constructed three visually identical Defender 110s for use in different filming locations. All were RHD open models with V8 engines and automatic gearboxes, fitted with special racking to look the part of the tough adventure vehicle. The company also supplied a tough-looking Freelander for the film, but in the event this was not used.

Capitalizing on the Defenders' appearance in the film, Land Rover prepared a special edition. The vehicles that came off

The Tomb Raider special edition came as either a 90 Station Wagon or a 110 Double Cab.

the Solihull assembly lines did not look very much like the film 'specials', but there were similarities in their Bonatti Grey paint, chequer-plate body protection panels, rollover cage with roof rack and roof-mounted spotlights. Even the interior detailing owed something to the film vehicles. There were also large circular Tomb Raider badges as standard and three different Accessory Packs were available to individualize the vehicles.

Defender Tomb Raider (Australia)

Land Rover Australia kept costs down by making its own Tomb Raider limited edition out of Defender 110 Station Wagons with the local Xtreme package. The limited-edition models were created by dealers, who bought a kit of parts from LRA and fitted it themselves. Exactly thirty kits of parts were sold between June 2001 and April 2003. In theory, they were to be fitted to Bonatti Grey vehicles, but there is no guarantee that some of the kits did not end up on 110 Station Wagons of other colours.

The kits of parts consisted of a bull bar, side rails, a roof rack with twin spotlamps and a ladder, bonnet and wing-top protector plates, chequer-plate lower body protection panels, headlamp guards, a sump guard and a raised air intake. There was no special badge of any kind.

The 2002 Model Year

Defender 90 Arusha (France)

Land Rover France introduced the Arusha special edition at the Val d'Isère 4×4 show that opened on 11 August 2001. The model was a French-specification hardtop (i.e. a Station Wagon with no rear seats) and went on sale at the start of the 2002 model-year. It is not clear how many were sold.

Its name came from the city in northern Tanzania, and was intended to convey an idea of adventure. This was backed up by a zebra-stripe finish (actually white paint with black decals). There were black wheel-arch eyebrows, four mud flaps and a tow hitch as standard.

There were three seats in the driving compartment, upholstered in Techno cloth, and the Arusha had a Roof Appointment Pack with Station Wagon-style headlining. Perimetric and volumetric alarms were also fitted.

Defender 90 Braemar (Scotland)

The success of the X-Tech limited-edition commercials inspired Land Rover to create another similar edition for sale only through its dealers in Scotland. The Defender 90 Braemar was introduced at the Scottish Motor Show in November

The Braemar was a regional special edition for the commercial market in Scotland.

ABOVE: **Large decals characterized the Sun Rider limited-edition cabriolet for France.**

LEFT: **The angle makes it hard to tell, but this Defender Black is in fact a 110 Double Cab version.**

2001 and consisted of twenty-five hardtop models in Blenheim Silver – so these were in fact 2001-specification models despite their release date.

The Braemar models also had County-style cloth seats, tinted glass and a sunroof. The Freestyle Choice wheel, tyre and anti-roll bar package was standard. There were side runners and a rear step, front mudflaps, spare wheel cover and heated rear window with wash-wipe. Drawn from the options list and standard on the Braemar were a raised air intake, an A-frame soft front protection bar, Rally 1000 driving lamps with stone guards, a swift-access ladder, chequer-plate on the body and an M6000 winch.

Defender 90 Sun Rider (France)

For summer 2002, Land Rover France introduced yet another special-edition cabriolet with the Challenge 4×4 conversion. There were forty-five examples of this, all of them for France and all of them with special Sun Rider decals on the doors.

Defender Black (UK)

The Defender Black, claimed Land Rover dealer literature, was a 'new Defender for the urban warrior'. The description was perhaps rather cynical, but it did reflect the fashion of the time for urban dwellers with an active lifestyle to drive tough-looking 4×4 vehicles. To make that point, Land Rover announced the new model at the Canary Wharf International Expo in London, which opened on 10 June 2002.

There were 250 Defender Black models, of which 100 were 90 Station Wagons and the other 150 110 Double Cabs. All of them, as the name suggested, were painted black. They came with Boost alloy wheels, chequer-plate panels on the wing tops and body sides and side protection rails. The upholstery was in leather, with a leather-bound steering wheel, chrome detailing on the instrument bezels and aluminium alloy grips for the gear and transfer box levers. Air-conditioning and a single-slot CD player were standard, as was the ABS-plus-ETC package.

The 2003 Model Year

Defender 90 Technium (France)

The French Technium edition was introduced in August 2002 at the Val d'Isère 4×4 show as a 2003 model. It was essentially a Defender 90 Td5 in Zambezi Silver with County trim and an 'aluminium' theme. There were ninety examples altogether; seventy were French-specification hardtops (Station Wagons with no rear seats) and the other twenty were full Station Wagons. All had heated rear windows.

There were Technium logos on the centre pillars and tail panel, chequer-plate wing protection panels and the rear lamps featured a round fog lamp on the left and a round reversing lamp on the right. Front mudflaps, a tow hitch and bumper end caps were standard and the wheels were Boost alloys with 235/85R16 tyres. Both main and transfer gear levers had aluminium grips, the radio came with a CD player and both a perimetric alarm and an immobilizer were fitted. Air-conditioning was an extra-cost option.

The French Technium edition had an aluminium theme.

Defender 90 Highlander (France)

The Highlander was a French special edition introduced in approximately May 2003. It came as a soft top (without side windows in the canvas), a French-style hardtop (Station Wagon with no rear seats) or a Station Wagon, and there were 120 examples in all.

Probably all the County colours were available and the Highlander had body-colour wheel-arch eyebrows and Highlander logos on the front wings and tail panel. The wheels were Boost alloys with 235/85R16 General Grabber tyres and a tow hitch was standard. Cloth upholstery was standard, as were electric windows, remote central locking and a radio with CD player. Air-conditioning was an extra-cost option.

The 2004 Model Year

Defender 110 Black (Brazil)

The Brazilian assembly plant produced a limited run of 110 Black models at the start of the 2004 model year. These were based on the Station Wagon and featured dished five-spoke alloy wheels with a gunmetal finish, chequer-plate on the sills and rear lower body, and side runners with a chequer-plate step.

Defender G4 Edition

The Defender G4 Edition was marketed in several countries, taking its cue from the 2003 Land Rover G4 Challenge sports adventure event. All models were Station Wagons and there were both 90 and 110 types. The four paint colours available were Tangiers Orange (the colour associated with the vehicles actually used on the G4 Challenge), Borrego Yellow, Java Black and Zambezi Silver. Quantities varied from one territory to the next. The UK, for example, had 150 vehicles in all, while France had fifty-seven, consisting of thirty-eight short-wheelbase and nineteen long-wheelbase models.

All variants came with Boost alloy wheels, an external roll-over bar around the cab, black chequer-plate on the sills and wing tops, front and rear lamp guards, an A-frame front protection bar and a rectangular identifying decal on each front wing. The interior featured Black Mogul fabric seats with PVC side bolsters, a Foundry finish switch panel with a radio incorporating a single-slot CD player and air-conditioning.

The Brazilian CKD plant also built its own special edition. The Defender Black was based on what they called a Standard model – a 110 Station Wagon – which explains that odd rear decal.

The G4 Edition tied in with the first Land Rover G4 Challenge.

Defender 90 X-Tech (UK)

From November 2003 there was a third X-Tech edition based on the Defender 90 County hardtop. This time, there were 100 examples, built in October and November 2003. The format broadly followed the previous editions in 2000 and 2001, but with changes.

The vehicles were painted in Zambezi Silver, with body-colour roof and wheel-arch eyebrows, plus a Brunel grille and lamp surrounds as used on the XS models. There were smoked lenses for the front turn indicators and side repeaters, and side runners with chequer-plate steps. The Freestyle Choice package of Boost alloy wheels with 235/85R16 tyres and anti-roll bars was standard.

The cabin was trimmed completely in black, with leather and Ripstop upholstery, two seats and a centre cubby box. Leather binding was used for the steering wheel rim and the main and transfer gear-shift grips. The dashboard switch panel, however, was in Brunel silver; air-conditioning, the Cold Climate Pack (heated seats and windscreen) and a single-slot CD player with four speakers were all standard. There was also a stowage net inside the tail door. An optional Protection Pack brought chequer-plate wing top and sill protection panels for extra cost.

The 2005 Model Year

Defender 90 Libyan Sand (France)

Land Rover France's usual late-summer special edition for September 2004 was related to the German-market Experience special edition launched at the same time. Both had the same special Libyan Matt Sand paint (coded LRC133).

There were 150 examples of the Defender 90 Libyan Sand, of which seventy-five were French-specification hardtops (Station Wagons with no rear seats) and seventy-five full Station Wagons. The Libyan Matt Sand paint was matched by

The French Libyan Sand had a special body colour and special decals as well.

white steel disc wheels and white roofs; there were black wheel-arch eyebrows, black side protection bars and dark-finish chequer-plate protection panels on the wings. A tow hitch was standard.

Other identifying features were graduated stripe decals around the leading edges of the rear side windows. Within each of these and on the tail panel was a circular graphic with the Libyan Sand name and a silhouette of a man with a camel.

Defender 110 Experience (Germany)

Land Rover Germany used a special edition of the Defender 110 to promote its Experience off-road driving tuition centres in September 2004. There were 150 of these models, painted in the same Libyan Matt Sand as the French Libyan Sand edition that was released at the same time.

Unlike the French vehicles, these 110s had their steel wheels and roof in the body colour. They had black wheel-arch eyebrows, black chequer-plate front wing-protection panels and a snorkel on the right-hand front wing. A rear tow hitch was standard and each front door carried a large Land Rover Experience decal logo. The interior featured heated front seats and a CD player with four speakers. The ABS-plus-ETC package was included in the specification.

The cost of 34,500€ included a one-day off-road driving course at a Land Rover Experience Centre.

Defender Silva (Belgium and Holland)

The Defender Silva was a co-branding exercise between Land Rover and the Swedish Silva company, manufacturer of outdoor gear. It was available in Belgium and in Holland as a 90 hardtop (Station Wagon with no rear seats), a 90 Station Wagon or a 110 Station Wagon. No quantities are known.

The Silva edition was probably released in January 2005. It was available in several metallic and micatallic colours, but the promoted colour was Zambezi Silver. Roofs and wheel-arch eyebrows were in the body colour. To suit the image of outdoor adventure, Silva models came with an exterior roll cage over the cab, chequer-plate on the wings and sills, a snorkel air intake, lower bash plate and protective grilles over the front and rear lights. There were Boost alloy wheels with 235/85R16 General Grabber tyres and four mudflaps.

All glass was tinted and there were electric front windows. The seats had cloth upholstery, the radio-cassette came with two speakers and an alarm was standard. The vehicles were branded with a Silva logo on the spare wheel cover and came with a Silva compass and keyring.

Defender Black Silver (France)

Land Rover France hedged its bets on paint colours with the Black Silver edition, announced in March 2005. There were 220 examples in all: 100 were 90 Station Wagons; sixty were 90 hardtops (Station Wagons with no rear seats); forty were 110 Station Wagons; and twenty were 110 Double Cabs. In each case, half the vehicles were painted Zambezi Silver and the other half Java Black. Chequer-plate protection panels on the wings and sills were in silver with black bodywork, or black with silver bodywork.

All variants had Boost alloy wheels, a 'soft' front-protection bar and a pair of long-range driving lamps. There were 'Black Silver' badges on the sides and tail, while the Double Cab models came with a tonneau cover and rear-mounted spare wheel carrier. All interiors had a Brunel silver switch panel, a Sony CD and MP3 player, plus aluminium grips for the main and transfer gear levers.

The G-Star Raw Defender 110 (Netherlands)

The G-Star Raw Defender resulted from a co-branding exercise between Land Rover and the denim clothes brand G-Star in Holland. It was announced in July 2005 with 'raw aesthetics' as its theme, but it is not clear how many were made. Initial publicity promised that the vehicle could be made available in other European countries if there was interest, but it seems likely that the G-Star Raw Defender actually remained a unique vehicle.

The base vehicle was a Defender 110 Station Wagon, painted in matt Pennine Grey with a brown leathercloth roof. Grille, bumper, undershield and wing vents all had punched holes, and the brown spare wheel cover carried the G-Star logo. The door handles and Boost alloy wheels were finished in the body colour. The vehicle was badged as a RAW Defender on the bonnet, front wings and rear body sides, and there was a numbered limited-edition plate riveted to the tail panel. The first vehicle was numbered 3301 (the G-Star production code for its denim cloth).

Upholstery was, unsurprisingly, in denim with the number 3301 emblazoned across the seat backs. The vehicle had air conditioning, a CD player, tinted glass, electric front windows, heated seats, a tow hitch and an alarm. It also had the standard European features of luggage hooks in the rear and the ABS-plus-ETC package.

Defender 90 Bel Air (Belgium and Luxembourg)

Land Rover's national sales companies in Belgium and Luxembourg shared a limited-edition cabriolet model from July 2005. The Bel Air was painted Java Black and came with the cabriolet conversion manufactured by Challenge 4×4 in France; the cabriolet hood was black.

There were seventy-five examples of the Bel Air. They had Boost alloy wheels with General Grabber 235/85R16 tyres and Bel Air logos on the front wings and tail panel. There were no Td5 logos on the front wings and only rear mudflaps were fitted.

The 2006 Model Year

Defender Silver (UK)

This UK special edition launched in September 2005 had a number of similarities to the silver models of the French Black and Silver edition available earlier in the year. There were 300 examples in all, and the models available were a 90 Station Wagon, 110 Double Cab and 110 Station Wagon.

The Zambezi Silver bodywork had matching wheel-arch eyebrows and roof, with contrasting dark chequer-plate on sills, rear lower corners and wing tops. There was an A-bar with long-range lamps, an undershield, a tow hitch and grilles over the front and rear lights. The wheels were Boost alloys.

The Raw Defender was the first to have matt paintwork, at a time when it was just becoming fashionable.

The interior featured tinted glass, electric windows and black seats upholstered in Ripstop cloth and leather. A CD player and central locking completed the specification.

Defender Limited (Germany)

This German special edition for 2006 was a further reminder that black and silver were the colours most in vogue at the time in Europe. The focus of this one was on urban sophistication. The Limited was announced in September 2005, but build quantities are not known.

There were both 90 and 110 Station Wagons, all in Java Black with Boost alloy wheels. Roof and wheel-arch eyebrows were in black, and there were black side-runners and a spare wheel cover with Land Rover logo. Only rear mudflaps were fitted.

The interior had special Cognac leather upholstery with the Land Rover logo on the backrests of the front seats, which were heated. The interior door handles and the grips of the main and transfer gear levers were in aluminium and there was a numbered limited-edition plate on the dashboard. A CD player was standard, but air conditioning and the ABS-plus-ETC package were both extra-cost options.

Defender Black Edition (Switzerland)

Released in April 2006, this limited edition for Switzerland was once again in Java Black. There were forty examples, divided between 90 and 110 Station Wagons. The Black Edition had similarities to both the French Black and Silver and to the German Limited models.

Roof and wheel-arch eyebrows were in the body colour and the grille and headlamp surround panels were in XS-style Brunel silver. There was silver chequer-plate on the sills and rear lower body corners, plus a Black Edition logo on each rear wing. The wheels were Boost alloys and there was a soft front A-bar with long-range driving lamps and stone guards, plus side runners. There were clear lenses for the front and repeater indicator lights.

The upholstery was in brown Nappa leather with the Land Rover logo on the backrests of the front seats. The front seats and windscreen were heated. Interior door handles and the two shift grips were in aluminium, the steering wheel was leather-bound and there was a numbered limited-edition plaque. Air-conditioning was standard and all Black Edition models came with a TomTom Go 700 sat-nav with touch screen. The radio-CD unit was Bluetooth-equipped. Both 90 and 110 Station Wagons had four individual folding rear seats and a luggage net on the rear door.

Defender G4 Challenge (France)

This French special edition was released to benefit from the publicity associated with the second G4 Challenge event and reached showrooms in April 2006. There were 115 examples, all based on the Defender 110 Station Wagon.

Three different colours were available: Tangiers Orange (the special G4 Challenge colour), Java Black and Zermatt Silver, in each case with body-colour roof but black wheel-arch eyebrows. Each front wing carried a 'G4 Edition' decal, identical to that used on the 2003 global G4 Edition. There were chequer-plate protection panels on the wings, sills and lower rear body, plus black side runners, a snorkel air intake and a Land Rover branded roof rack. The wheels were Boost alloys and there was a soft A-bar for frontal protection.

The interior featured Ripstop and leather upholstery in black, air-conditioning, rubber overmats and a rubber load area mat, plus an RDS radio with CD player.

Defender 110 Experience (Germany)

The first of the three German limited editions released in spring 2006 was the Experience, a Defender 110 Station Wagon in Zermatt Silver with the Land Rover Experience logo on each front door. It resumed the theme of the 2004 Defender Experience special edition and there were fifty examples, with sales beginning in April 2006.

Defender Style (Germany)

There were 150 examples of the Style limited edition, which became available in Germany in May 2006. The edition consisted of both 90 and 110 Station Wagons, all in Java Black paint with a Brunel silver grille and headlamp panels, Boost alloy wheels, air-conditioning and a radio with CD player as standard.

Defender 90 Cabriolet Two (Germany)

Also released in May 2006 was the Cabriolet Two edition of 100 vehicles based on the Defender 90 with Challenge 4×4's cabriolet conversion. The 'two' in the name referred to the fact that there were just two seats, which of course were separated by a cubby box.

One of these vehicles was turned into a Cabriolet Luis Trenker by the sports fashion company of the same name and was exhibited at the Ispovision show in Munich in July 2006. The vehicle was the work of Luis Trenker's chief designer, Hugo Valentino. He wanted to display the company's trademark colours of red, black and white, and had the vehicle finished in white, with white steel wheels and red and black flower decals.

The German Cabriolet Two was considerably more sober than the French special-edition cabriolets – until Luis Trenker's chief designer got his hands on one.

The 2008 Model Year

Defender SVX

The Defender SVX was a special edition created for the 2008 sixtieth anniversary and revealed to the public at the Geneva Motor Show on 4 March 2008. However, the prototype had been seen several months earlier, when it had been unveiled by Zara Phillips, one of Land Rover's celebrity Ambassadors and grand-daughter of the Queen, at the British Red Cross Ball on 22 November 2007. That vehicle was donated to the Red Cross for sale by auction at the Ball.

In the UK, the SVX came only as a stylish two-seater soft-top or a 90 Station Wagon, but there were 110 Station Wagons as well in some countries. This edition was also the first to have special Recaro front seats.

There were 1,800 SVX models for Land Rover's markets around the world; of those, 300 were for the UK. However, the UK received only 90 soft-top and 90 Station Wagon versions; some other countries also received an SVX 110 Station Wagon. The dates of availability also varied; in the UK, soft-top models reached the showrooms in early summer 2008, but the Station Wagons did not arrive until late summer.

All the SVX models were painted in Java Black, with matt black side decals featuring a cut-out of the '60 Yrs' logo, plus plate-type badges with the SVX name on the front wings and tail panel. They had new and bolder grille and headlight panels, finished in contrasting silver, silver-finish tubular side runners and there was a matching silver-finish tubular rollover cage for the soft-top models. Headlamps had a new clear-lens design and were backed up by projecting side and indicator lamps, while the rear lamps had a new LED design and clear lenses. The wheels were a new five-spoke design with a two-tone finish, diamond-turned for the main part, but with the name Defender cast into one spoke and given a grey background, while the same grey was used for the wheel centres. A reinforced aluminium front undershield was also fitted.

Interiors all featured custom-made Recaro bucket seats at the front, with Land Rover logos on their integral head restraints. The shift levers had alloy grips and there was a new audio system with a subwoofer and USB socket, plus an iPod cradle. A Garmin satellite navigation system was also included and there was a numbered limited-edition plaque. Soft-top models had the spare wheel mounted vertically behind the seats and a flat cargo space behind that. They also came with a specially tailored removable black fabric hood.

Defender Sixtieth Anniversary (Switzerland)

In Switzerland, Land Rover marketed some special offers on most models during 1998 in connection with the sixtieth anniversary. For Defender, the special offers amounted to high equipment levels at no extra cost and the sixtieth anniversary, or Celebration, models were available as 90s, 110s or 130s. Quantities are not known.

The theme colour was Zermatt Silver, although other colours may have been available. The vehicles were more or less to SE (County) specification, with roofs and wheel-arch eyebrows in the body colour. Boost alloy wheels were standard, as was tinted glass with electric front windows and cloth upholstery. Carpets, a CD player and remote central locking were all standard and the Winter Pack of heated front seats and heated windscreen was also fitted.

Defender 110 Edition 60 Years (Germany)

Land Rover Germany also marketed its own special anniversary variant of the 110 Station Wagon. These had some similarity to the 2006 German limited special edition.

There were 150 of them, all in Java Black with body-colour roofs and wheel-arch eyebrows. There were black side runners, but the grille and headlamp surrounds had a contrasting Brunel silver finish, the side air intakes were in silver and there was silver chequer plate on the wing tops. The front wings carried plates reading 'Edition 60 Yrs' and there were shadow-black decals and '60' logos on doors and bonnet.

Leather upholstery had gone down well on the Defender Limited model and there was leather upholstery in the Edition 60 Years as well, this time in beige with black piping and Land Rover logos on the backrests. Air-conditioning, a centre cubby box and the Winter Pack of heated front seats and windscreen were all standard.

The 2009 Model Year

Defender Fire and Ice

The Fire and Ice edition was announced in July 2009, during a period of global recession. It was intended to boost sales during this difficult time, but was sold only in specific markets outside the UK where Land Rover believed that a more adventurous appearance would have an appeal. There were 850 examples in all. These were sold in Austria, Belgium, Brazil, China, France, Germany, Hong Kong, Hungary, Italy, the Netherlands, Poland, Russia, Singapore, South Africa, Spain and the Ukraine. Individual totals are not all known, but there were seventy-five for the Benelux (Belgium and Luxembourg) countries and 100 for France.

All models were Station Wagons with ABS and ETC as standard. However, the key feature of this special edition was its eye-catching multicoloured exterior. The main body colour on 90 models was Vesuvius Orange ('Fire') and that on the 100s was Alaska White ('Ice'). Vesuvius Orange was the colour used on the 2004 Range Stormer concept car. In

The Fire and Ice edition was a bit bright for UK tastes, especially in the middle of an economic downturn, but it went down well elsewhere.

both cases, Santorini Black was used for contrasting sections – the roof, the wheel-arch eyebrows, the side door hinges, the bonnet, the grille and the wheel centres. The grille and light surrounds were the protruding type seen on the 2008 SVX, while the wheels were special five-spoke alloys with the Defender name cast into one of the spokes. The Land Rover oval badges on grille and tail panel had white backgrounds and there was a special cast Defender nameplate on the tail panel. Black side runners were standard.

There were more special features inside. Recaro bucket-type front seats were fitted and the upholstery was in Alcantara with leather bolsters and a Land Rover logo on

Just ten of these soft-top models were built for Dutch customers. The grille was the same as on the 1999 Heritage edition.

the head restraints; the centre cubby box was upholstered to match. The front seats were heated and there were special branded Defender carpets. The faceplate of the centre switch panel was finished to match the exterior paint and the steering wheel was bound in leather, with stitching again matching the exterior colour.

The 2010 Model Year

Defender 90 Classic (Holland)

This was a very limited edition indeed, announced in March 2010. It consisted of just ten examples, all Defender 90 soft-tops in Keswick Green. Distinguishing features were a fake mesh grille like that used on the Heritage Edition in 1999, a beige soft-top, gloss black wheel-arch eyebrows and steel wheels finished in Broken White.

Defender Cap Horn (France)

Land Rover France announced the Cap Horn edition in April 2010, consisting of 100 vehicles. There were both Defender 90 hardtops (Station Wagons with no rear seats) and Defender 110 Station Wagons, in either Santorini Black or Stornoway Grey with body-colour roof and wheel-arch eyebrows.

The wing tops, sills and lower body corners had chequer-plate protection panels, in silver with black paint or in black with grey, and the wheels were Boost alloys with a two-tone finish. There were special brightly coloured Cap Horn plate logos.

Electric windows and a centre cubby box were standard and the Cap Horn models came with a CD player and a Garmin Nuvi 1390T satellite navigation system with Bluetooth compatibility.

Defender 110 Experience Limited Edition (Germany)

Land Rover Germany indulged its customers' liking for all-black 110 Station Wagons yet again with the Experience Limited Edition in 2010. The vehicles came with the ABS-plus-ETC package as standard. Quantities are not known.

The vehicles were painted Santorini Black with body-colour roof and wheel-arch eyebrows and black chequer-plate on the front wings. They had diamond-finish five-spoke alloy wheels with dark centre sections and a dark background on the spoke bearing the Defender name. All examples had a front undershield, black side runners, a soft front A-bar and an Experience plate badge on the tail panel.

The interior was characterized by Recaro front bucket seats flanking a cubby box, a three-spoke sports steering wheel with silver highlights, aluminium grips for the main and transfer box shift levers and air-conditioning as standard.

The Defender Collection (France)

This 2010 special promotion from Land Rover France was not a limited edition but a set of bright decals that could be added by dealers to any 90 hardtop (Station Wagon with no rear seats), 90 Station Wagon or 110 Station Wagon. The decals could be added at the point of sale, or to a customer's own vehicle.

These flowery decals from the French 'Defender Collection' were supposed to evoke the 1970s – but actually recalled the late 1960s.

The decals ran the full length of the vehicle's flanks and in some cases encroached on the roof as well. All were made in 50-micron vinyl that was claimed to last for more than ten years. There were five different types available, representing themes of the 1970s, the City, the Jungle, Nature and Travel.

The 2011 Model Year

Defender Rough

The success of the Fire and Ice edition in 2009 inspired Land Rover to produce another special edition with several similarities. The Defender Rough was displayed at the Brussels Motor Show in January 2011 and was available in Austria, Belgium, the Czech Republic, France, Germany, Holland, Italy, Norway, Poland and Switzerland, but not in the UK. The Defender Rough came as a 90 Station Wagon or 110 Station Wagon, in Nara Bronze or Zermatt Silver and some countries took 'hardtop' versions of the 90 Station Wagon without rear seats to qualify for commercial-vehicle tax discounts. All models had the ABS-plus-ETC package.

Both models had the SVX-style grille and headlamp configuration. The headlamp surround panels, but not the grilles, were in black and there were matching black roofs and wheel-arch eyebrows. There were also black enamelled alloy wheels with a new saw-tooth ten-spoke design and Goodyear MT/R tyres. The Land Rover name on the bonnet was in gloss black, but the oval logos on the grille and tail panel had the Titan (white background) style. The Defender decal on the tail panel was in black capital letters. The rear lights were LED types and there was a NAS-style rear step bumper.

Upholstery was in cloth with leather bolsters; Almond cloth was used for Nara Bronze vehicles and Ebony for silver ones, in each case with Ebony leather. There were Land Rover logos on the front head restraints. The cubby box had part-leather upholstery and the steering wheel was leather-bound with stitching in the exterior colour. The centre switch panel was also in the exterior colour. Air-conditioning and the Winter Pack of heated seats and windscreen were both standard.

Defender 90 X-Tech (UK)

The X-Tech series of limited editions for the UK had always been for the commercial market and had always been silver

The Defender Rough edition was first with the new saw-tooth alloy wheels.

90 hardtops. The 2011 edition, announced at the Brussels Motor Show in January 2011, followed that tradition but was closely related in most other respects to the short-wheelbase versions of the Defender Rough that was seen at the same show. Land Rover referred to its 'raw and edgy style' – and also issued an interior picture to the media that showed not the X-Tech, but the Defender 90 Rough Station Wagon.

The 2011 X-Tech models went on sale in March 2011, with 150 being built. All were painted in Zermatt Silver with Santorini Black roof and wheel-arch eyebrows, and gloss black saw-tooth alloy wheels – the first time these had been available in the UK. The seats were upholstered in Ebony part-leather trim.

The 2012 Model Year

Defender Blaser Edition (Germany)

The German Blaser Edition was a co-marketing exercise with Blaser, the maker of sporting guns and shooting accessories. It was introduced on 31 January 2012 at the Jagd und Hund (Hunting and Dog) exhibition in Dortmund and the build quantity was to be sixty units.

The Blaser Edition was based on a 110 Station Wagon in Keswick Green with the SE (European XS) specification and was kitted out as a hunting vehicle. There were Blaser logos on the front doors and spare wheel cover, and the vehicles each had a roof rack, tow hitch, snorkel air intake, grille guard, electric winch and rear spotlight. The seats were upholstered in black leather with green stitching and had the Argali logo of Blaser's clothing accessories embroidered on the head restraints. A special feature was a custom-fitted gun case in the load area. This could carry up to five guns and had two drawers underneath to carry ammunition and other equipment.

Defender 110 by Piet Boon (Netherlands)

Dutch fashion designer Piet Boon announced a special 'Defender 110 by Piet Boon' on 9 May 2012. The vehicle was finished in matt grey metallic (called PB Metallic) with matt black grille and wheels, with a white roof. The interior featured Recaro front seats and quilted leather upholstery in Elephant Grey.

The Defender 110 by Piet Boon was a one-off, but there were plans to build as many as ten if orders were received.

RIGHT: **The 2011 X-Tech special edition shared many of its features with the Defender Rough edition that became available at the same time.**

BELOW AND BOTTOM: **Germany's 2012 Blaser Edition was designed with a hunting theme.**

ABOVE AND LEFT: **The Piet Boon Defender capitalized on the vogue for matt paintwork, but combined it with diamond-quilted leather upholstery.**

Defender 90 Eden Park (France)

Land Rover France produced another special-edition for summer 2012 in co-operation with the sportswear manufacturer Eden Park. The vehicles were actually assembled by specialist vehicle body builder Durisotti at Sallaumines, using a new soft-top conversion kit. Exact numbers were not announced when the vehicles were launched in July 2012.

Defender Exclusive (Belgium)

The Defender Exclusive was announced on 9 July 2012 for Belgium. It was in effect a repeat of 2011's Defender Rough edition, with the same options of Nara Bronze or Zermatt Silver, plus black saw-tooth wheels and gloss black roof and wheel-arch eyebrows. This time, of course, it was powered by the 2.2-litre Puma engine rather than the 2.4-litre type that had powered the earlier edition. Intended volumes were not quoted at the announcement.

Defender X-Tech (UK)

The 2012 Defender X-Tech was announced at the same time as the Defender Exclusive, on 9 July 2012, and was a special edition for the commercial market in the UK. However, whereas all previous X-Tech models had been based on the Defender 90, this one was available as both a 90 hardtop and a 110 Utility Station Wagon.

In all cosmetic respects, the X-Tech was the same as the Exclusive, which was itself not sold in the UK. No production volumes were given when the model was announced.

The 2013 model-year

65th Anniversary Edition

Land Rover expected to build a low-volume Defender commemorative edition for its 65th anniversary year in 2013, but neither the name nor the details of the edition had been announced as this book went to press. The actual anniversary was in April 2013, but the probability was that the special edition would not appear until July at the earliest.

The 2012 X-Tech edition for the UK was essentially a commercial-market version of the contemporary Defender Exclusive for Belgium. This was the first time a 110 commercial derivative had been part of a special edition.

CHAPTER ELEVEN

THE MILITARY DERIVATIVES

Military sales had been a very important element of the Land Rover business since the marque's earliest days and the One Ten and Defender families carried on the tradition. As Chapter 3 makes clear, potential military orders took up a lot of the Solihull engineers' time during the development stages of the coil-sprung Land Rovers. Military sales would go on to account for large numbers of the coil-sprung models built, too.

Land Rover policy is not to comment on overseas military sales, because its customers prefer not to have details of their orders made available to potential enemies. Nevertheless, the physical evidence of vehicles out in the field usually gives the game away sooner or later, even if precise details of the vehicle types and order size are not readily available. There are also some military users who have been quite happy for at least the broad details of their Land Rover acquisitions to be made public and among them have been the British Armed Forces.

Land Rovers had been the standard UK front-line vehicle in their class since 1956 and as the One Ten came on-stream during 1983 there were still uncompleted orders for Series III 109in (2,769mm) and Series III 88in (2,235mm) Half-Ton ('Lightweight') models. Deliveries of these vehicles would not be completed until 1984–5, which gave some breathing space for Land Rover to settle a detailed specification for deliveries of the new coil-sprung vehicles. In the meantime, the first military order for One Tens was placed by the Singapore Ministry of Defence during 1984. It was for a fleet of 900 vehicles – an excellent start to the One Ten family's military career.

This very early military demonstrator showed the One Ten HCPU equipped with a missile launcher. Note the old-style Land Rover badge, perhaps fitted because no proper One Ten badges were available when this vehicle was being prepared.

This was a One Ten military demonstrator from the 1980s, with the twin fuel-filler arrangement; the standard rear tank was supplemented by one under the right-hand front seat.

This is one of the very first military One Tens, pictured at Solihull before delivery to the UK MoD for trials. It has the side lockers and full soft-top characteristic of the military specification, plus towing eyes on the front bumper. The aerial support ahead of the wheel arch and tuner box support on the front wing show that this is an FFR model. The galvanized windscreen surround and twin fuel fillers were not carried over to the UK fleet orders.

Early UK Military Deliveries

The UK Ministry of Defence requested a trials batch of One Tens before placing its first order. The new military designation for the class of vehicle into which the One Ten fell was TUM, which stood for Truck, Utility (Medium), and therefore the trials were conducted under this name.

There were about ten vehicles in that trials batch, all One Ten soft-tops with the 2.25-litre petrol engine and including RHD, LHD, 12V GS (General Service) and 24V FFR (Fitted For Radio) specifications. They were ready by May 1983, just a couple of months after the civilian models had been announced; the first ones were delivered to the Ashchurch vehicle distribution depot early that month. They were registered in the KB series used for all three services and for all vehicle types – these and later trials vehicles were the only One Tens in that registration series.

The trials vehicles were put through their paces on the military trials ground at Long Valley from about July 1983, as a result of which some elements in the specification were modified. The MoD wanted a single fuel tank at the rear instead of the twin-tank (rear and side) arrangement on the trials models. They also rejected the original galvanized windscreen frames in favour of painted frames. However, one feature new to British military Land Rovers that did get carried over to production was the provision of lidded lockers in the middle of the wheelbase.

By the time the first orders were delivered in 1985, the MoD was standardizing on diesel engines and its One Tens came with the latest 2.5-litre naturally aspirated type. There were some problems with these in the early days, as drivers and mechanics alike had to get used to the requirements of diesel engines that were very different from the petrol engines in older Land Rovers. The engine breathers gave trouble, too, but the issue was soon resolved.

Like most subsequent military deliveries, these first MoD One Tens had doors with removable tops containing sliding windows. As Chapter 4 shows, the civilian specification had moved on to one-piece doors with winding windows and later still would move on to incorporate car-like exterior door handles. However, the standard military specification always retained these two-piece doors with their recessed outer handles. Though their configuration was traditional to Land Rover, they became known as 'military-specification' doors to distinguish them from the one-piece type used on standard production models.

Most of these early One Tens for the UK MoD were soft-tops, although there were also some hardtop derivatives, mainly used for communications duties. In addition, the Royal Navy bought a batch of One Ten ambulances in 1986.

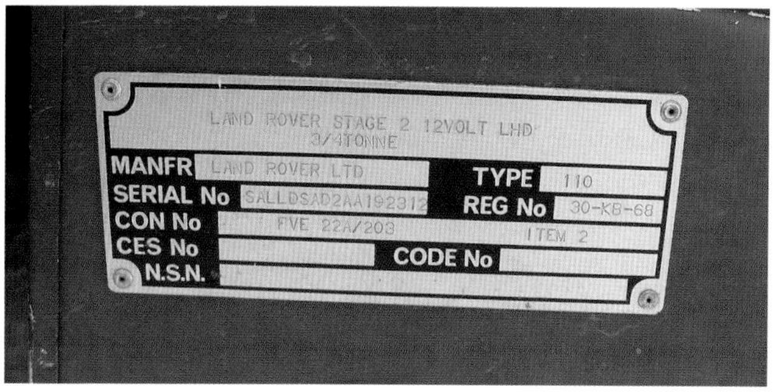

All British military Land Rovers carry a military identification plate on the seat box similar to this one, but this plate is special. It is on one of the original UK military trials vehicles from 1983. The 'Stage 2' description would not be used for the production fleet.

UK: The SAS One Ten V8

Those first UK trials vehicles were followed by others, including a One Ten with a V8 petrol engine and High-Capacity Pick-Up (HCPU) body, registered as 90 KB 62 in May 1984. This was the trials vehicle for a very special series of military Land Rovers – replacements for the long-serving Pink Panther long-range desert patrol vehicles used by the Special Air Service (SAS).

These British GS models were pictured when new in late 1985 or early 1986. They have the standard UK military lighting configuration, which dispensed with the headlamp surround panels to improve access for changing bulbs. Note also the towing pin in the centre of the bumper.

The SAS (which is able to bypass the normal and rather cumbersome MoD procurement system) lost no time in deciding what it wanted. Once satisfied with the basic One Ten V8 HCPU, the Regiment engaged specialists Glover Webb to design the long-range desert patrol specification that it needed. In fact, the SAS One Tens were designed as dual-purpose vehicles and much of their special equipment could be demounted so that they could be used as general-purpose load carriers, complete with a full-length canvas tilt. Whether this happened in practice is not clear.

The SAS One Ten V8s were some of the first One Tens to enter service with the MoD, the first batch of thirty-three arriving in November 1985. These were registered in the KD series that was then current and, although they were delivered from Glover Webb in standard NATO green with full windscreens, it was not long before they lost those windscreens and were repainted in sand camouflage. Despite the beige colour, they were once again known affectionately as 'Pinkies' like their Series IIA forebears.

That the SAS could risk using vehicles with the notoriously thirsty V8 engine for long-range work behind enemy lines was explained by changes in resupply logistics since the original Pink Panthers had been built. Those had been fitted with huge long-range fuel tanks which gave them a range of some 1,200 miles (1,930km). However, by the end of the 1970s, the SAS was using both helicopters and Mercedes-Benz Unimog 'mother ships' to deliver supplies to patrols in remote areas. So the 200-mile (320km) range of a heavily-laden One Ten V8 relying on military-pattern twin underseat fuel tanks was not going to be a serious handicap – and the extra power of the V8 engine was certainly welcome.

The SAS One Tens were equipped in service with 9.00 × 16 tyres to give additional ground clearance. They were designed for a crew of three and carried a forward-facing machine gun and a post-mounted machine gun in the rear. Safety considerations dictated a rollover bar behind the cab and a vertical grille or metal panel behind this was used to hang more equipment. The back body carried jerrycans as needed, ammunition boxes, additional weaponry for the operation in hand and had a covered 'cupboard' for food and other supplies. There were two spare wheels, one on the bonnet and one mounted on the side behind the driving compartment.

There was a front-mounted winch for self-recovery (a UK-made Superwinch 8552 in a specially made bumper). Pioneer tools, sand ladders and other paraphernalia were distributed around the flat surfaces of the body. All vehicles carried the then-new Magellan GPS equipment, plus a traditional sun compass mounted on the dashboard. Beyond that, what went on board was down to the crew. They knew what they needed for their mission and they packed it into their bergens or stuffed it into bags slung from every hook or bracket fitted to the vehicle. Camo netting went across the scuttle in the traditional way and more would be slung across the rear or carried within the back body.

The SAS One Ten V8s were not the only vehicles of their type to be built. Glover Webb built a batch of broadly similar

The SAS One Ten V8 long-range desert patrol vehicles were the successors of the famous Pink Panthers. This preserved example seems to be covered in equipment – much as the vehicles would have been in service. Note the 9.00 x 16 tyres.

LEFT: **The RAF Mountain Rescue teams used One Ten Station Wagons like this one, which entered service in late 1989.**

BELOW: **The Glover Webb Remote Area Patrol Vehicle was the basis of the SAS One Ten V8s. This example was a demonstrator, similar to the vehicles bought by the Jordanians.**

vehicles for the Jordanian armed forces as Remote Area Patrol Vehicles and these were also delivered during the 1980s. The SAS also took delivery of a further batch of One Ten Desert Patrol Vehicles, supposedly immediately after the Gulf War in about 1991. These vehicles were of course strictly Defender 110s and they also had the Tdi diesel engine instead of a petrol V8.

The MoD also began to take delivery of militarized Nineties from about 1986. Land Rover had of course made them aware of the new short-wheelbase model early on, but by the early 1980s there was less call for short-wheelbase models than in earlier times: the increased lifting capacity of the latest military helicopters allowed all units to make more use of One Tens. Although short-wheelbase vehicles still made good runabouts and liaison vehicles, there was certainly no requirement for a specially developed military variant such as the old 88in Lightweight. The Nineties were delivered to meet the requirements of what was now called the TUL, or Truck, Utility (Light) vehicle class. As with the One Ten, the initial contract was followed by a series of 'running' contracts that ensured regular deliveries of new vehicles to the MoD until well into the 1990s.

Australia: Project Perentie

Jaguar Rover Australia (JRA) reacted swiftly to hints about a replacement programme for the Australian Army light 4×4 vehicles, which were Series III models supplied between 1978 and 1981. The programme was to be called Project Perentie, after a large and fierce Australian lizard species. It looked very much as if the Army was going to make the switch to diesel power and the one thing that JRA could not offer at that stage was a diesel engine worthy of the name.

In fact, it was not only the military who wanted a decent diesel engine in its Land Rovers. Showroom sales of Series IIIs in Australia had been badly affected by the availability of a big 4.0-litre diesel in the Toyota Land Cruiser. So the planners at JRA came up with an idea that would kill two birds with one stone.

As there was no chance of a new diesel engine from Solihull in the short term, JRA pushed for early agreement to a bought-in engine. This would help showroom sales and provide experience of the engine well before the Army formally asked for tenders to replace its existing Land Rovers. This was doubly important, because the Army was known to be opposed to trialling prototype vehicles – it preferred vehicles that were already proven in the marketplace.

The JRA team looked at several options, choosing the 3.9-litre direct-injection engine made by Isuzu. It was a big, rough and rugged 4-cylinder that had been developed for eight-tonne trucks and certainly had enough power and torque to meet the needs and the likely requirement of a fifteen-year service life. JRA engineered one into a Series III 109 chassis and

Typical of the 4x4 Perentie models is this one. Note the special side lockers, the rear jerrycan holders and the low-slung rear lights.

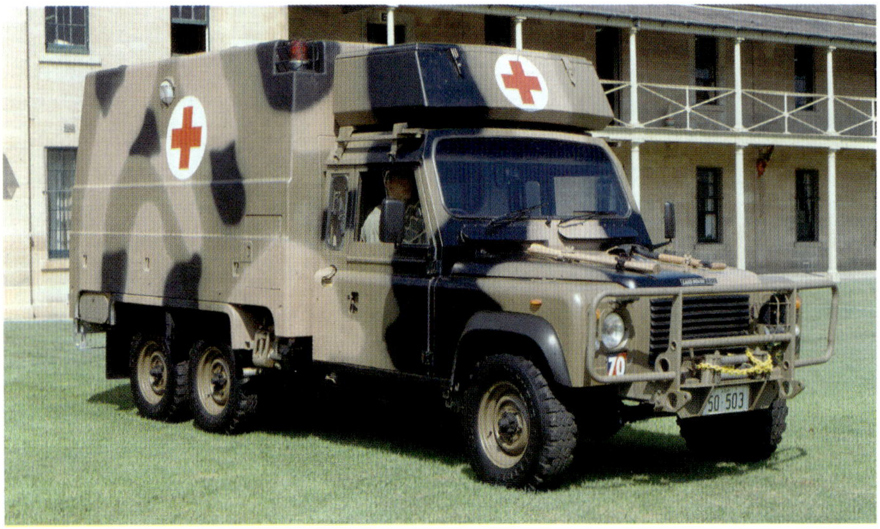

The Perentie 6x6 ambulance was one of several types to use the same basic body. On the ambulance, it was air-conditioned. The special wide cab, made of GRP, is obvious here.

Australian troops are seen here in Iraq, with a well-laden 6×6 Long Range Patrol Vehicle.

brought the combination to production in the Series III 109 as the 3.9D model in 1981.

It was also in 1981 that the Australian Army formally requested tenders for the new fleet of 1-tonne and 2-tonne vehicles it needed for delivery in the second half of the decade. The 1-tonne class was familiar Land Rover territory and JRA was sure that a suitably prepared long-wheelbase model would be a credible contender. The 2-tonne class was new territory, but it was clear that if JRA could come up with a 2-tonner that shared a high proportion of parts with its 1-tonne models, the overall package would be very attractive to the Army because it would simplify logistics support. If both the 1-tonne and 2-tonne vehicles could be built within Australia and could use a high proportion of local content, the package would be doubly attractive. Even better, there was a small but worthwhile civilian market for a 2-tonne truck in Australia.

So JRA decided to develop a 2-tonne Land Rover with the Isuzu diesel engine to meet both civilian and military requirements. The major work would be done in Australia, with support from Solihull as necessary, and in 1982 JRA appointed Ray Habgood as Engineering Manager to oversee the project. Several different ideas were discussed at this stage, including a UK proposal for an ultra-long-wheelbase One Ten. However, a 6×6 configuration seemed the best bet. The JRA team brought over a V8-engined Sandringham 6 conversion based on a coil-sprung One Ten prototype for evaluation, but decided they could do a better job themselves.

JRA began by making a 6×6 from a standard leaf-sprung 109 chassis with the Isuzu engine. A second development 6×6 followed it, still on leaf springs. These showed that simply extending the standard chassis would not deliver a vehicle rugged enough to meet the Australian Army requirement. The chassis side rails needed to be 50 per cent deeper in the centre to cope, which was 'totally impractical' in Ray Habgood's words. So a new chassis would be needed and the tooling costs for this would have to be minimal because of the relatively low build volume that JRA anticipated. This was why the new JRA 6×6 chassis ended up with its rear section made from square and rectangular steel sections welded together to form an exceptionally rigid platform.

This new rear chassis was mated to the front end of a One Ten chassis, bringing both disc brakes and coil springs into the specification for the third prototype. But the centre and rear axles were not coil-sprung. Unable to get the combination of load-carrying ability and ride quality that they wanted from coil springs, the JRA team had developed a remarkable load-sharing rear suspension using overlapping leaf springs attached to a rocker beam. It delivered exceptional axle articulation for off-road use, too.

As for a gearbox, the LT95 4-speed type was just right for the job. It was the strongest gearbox Land Rover then had available and it incorporated an additional rear output – designed for the powered-axle trailer that the British Army had at one time planned to use with the 101. This allowed a simple drive arrangement with a separate propshaft going to each of the rear axles. This in turn allowed for greater axle articulation than the through-drive axles used in 6×6 conversions like the Sandringham and also meant that cheaper, conventional rear axles could be used. However, development showed that the 3.54:1 gearing of the One Ten axles was too high for the 6×6 application, so the JRA team reverted to the 4.7:1 gearing of the old Series III models.

JRA's tender for the 1-tonne and 2-tonne military contracts was accepted and the company was asked to supply three 4×4 models and three 6×6 models for evaluation in September 1983. The 4×4s had the Isuzu-plus-LT95 power train and had also been extensively reworked from the standard One Ten specification in other areas. Their chassis side members were splayed outwards at the back so that the spare wheel could be accommodated under the load floor and there were modifications to the front end to accept a heavy-duty brush guard. The chassis frame incorporated reinforcements originally designed for the Shorland variant of the One Ten and there were a number of special features on the body.

As for local manufacture, JRA was already planning to set up its own assembly operation and to launch Australian-made One Tens into the civilian market in November 1984. The company invested some Aus $3 million into doubling the size of a plant in the Sydney suburb of Moorebank that British Leyland had been using as an emissions test and development centre; that investment was rewarded when they were awarded the contract to supply both the 1-tonne and 2-tonne vehicles to the Army. Signed in October 1986, the contract called for 2,500 1-tonne 4×4s and 400 2-tonne 6×6s to be supplied over the next four years. It was the largest single contract ever awarded to JRA and was worth in excess of Aus $150 million.

The Army had called for a series of changes and JRA duly beefed up the bearings in the LT95 gearbox – a modification also adopted in the UK for all variants of the gearbox. The axle tracks were also changed: on the trials vehicles these had been wider at the rear than at the front, but the production models had a single common track all round. All chassis were made of hot-dipped galvanized steel for longevity. But most important was that the Army had suggested a redesigned cab for the 6×6 models, the standard One Ten cab on the trials vehicles having proved too small for a crew of three soldiers over long distances.

The new cab, designed completely in Australia, was 200mm (7.8in) wider (accommodated by the wider tracks), 50mm (2in) taller and 65mm (2.5in) deeper, incorporating an opening vent window behind the seats and a special new single-piece windscreen that was locally made. Built on a steel spaceframe, it incorporated some standard Land Rover pressings, but did demand changes elsewhere in the vehicle. A new bonnet panel, a new grille panel and a new dashboard all had to be made up to suit the extra width; all were in fact manufactured from GRP. For cost reasons, however, the wide-cab vehicle was not made available on the civilian market and civilian 6×6s also kept the narrow-track axles with a disparity between front and rear.

The first Perentie models entered service on 6 February 1987 and the last on 21 January 1992. There were multiple variants of both the 4×4 and 6×6 models. The 4×4s came as soft-tops, some with winches and some FFR types, panel vans for survey duties, Station Wagons for commanders and as troop carriers, plus surveillance and reconnaissance variants. There were 6×6 cargo carriers (with and without winches), plus a range of modular 'van' bodies built of GRP on a steel framework by Jakab Industries in NSW that could, theoretically at least, be swapped from one vehicle to another. Variants were maintenance and electronics trucks, ambulances and (later) Comsec trucks. There were also five-man crew cab 6×6s, air-defence vehicles and, most readily recognizable of all, the Long-Range Patrol Vehicle (LRPV) produced for the Australian SAS Regiment (SASR).

The LRPV variants did not enter service until November 1991. There were just twenty-seven of them and they were designed to operate without logistics support behind enemy lines. In practice, many were also used in support of drug enforcement operations in Australia's Northern Territory. The LRPVs had long-range fuel tanks carrying an extra 80gal (35ltr) of petrol and were initially built with a combined rollover cage and weapons ring mount. They all also carried a 250cc Suzuki motorcycle slung across the back for solo reconnaissance missions.

When the Perentie contract had been awarded, JRA had known that there were likely to be further military orders and events proved them right. Unfortunately, JRA collapsed in 1992 and the Australian Land Rover operation became part of Rover Australia. As part of the restructuring of the company, the Moorebank CKD plant was closed down. There would be no more Isuzu-engined One Tens, no more 120s and no more 6×6s, either. So when the Australian Army called for some more 6×6s under Project Bushranger in the mid-1990s, a special arrangement had to be made with British Aerospace in Adelaide to have them manufactured there. The very last Bushranger 6×6s were thirty-seven flatbed types, which the Australian Defence Force equipped with the Parakeet satellite communications system.

There were some more military 6×6s for overseas armed forces, although the only one known for certain was for

LEFT: **The only known overseas orders for the Australian 6x6 was for logistics trucks for the Omani military.**

BELOW: **The variety of Australian military 6x6 models is illustrated here. All were procured under the Perentie contract.**

eleven 6×6 logistics vehicles for Oman. Land Rover in the UK did try to market the Australian 6×6 globally from 1989 and it was certainly offered both to the Canadians and to the Singapore Ministry of Defence (who wanted a short-chassis heli-portable model). However, neither of these potential customers placed an order. One military 1-tonne 4×4 and one military 2-tonne 6×6 were shipped across to the UK, where Solihull promptly handed them over to the MoD for assessment. However, nothing came of that venture, either.

The Perentie One Tens and 6×6s proved exceptionally successful in Australian military service, so much so that their intended service life of fifteen years was extended more than once. A plan to re-engine them with Td5 diesels in 1998 came to nothing (although a prototype was built), but the 6×6s in particular proved so essential that from 2004 they were put through an engineering programme at Tenix Defence of Bandiana, Victoria. This company was contracted to refurbish the vehicles to as-new condition in a contract that ended in 2010. It was a move that was expected to see the vehicles' service last until at least 2017. Though the contract for their replacements was awarded to Mercedes-Benz in 2008, the oldest Perenties were expected to have had a service life of thirty years by the time they were finally withdrawn.

One Two Sevens – Ambulances and Rapier Tractors for the UK

The One Two Seven chassis was also developed for military use and its first derivatives in British military service were a batch of Marshall-bodied ambulances. All of these had V8 petrol engines and the majority seem to have had registrations in the KG series. Some were configured specifically as Crash Rescue ambulances for the RAF, but others were intended for general ambulance duties by all three services.

The basic design was also available as a civilian ambulance (and the One Two Seven chassis of these vehicles were to civilian rather than military specification). The military versions were designed to accommodate a maximum of three stretchers, or two stretchers plus three seated patients. Their Luton heads were all integrated into the cab structure, there was a small round spotlight mounted on each windscreen pillar and the front bumper carried a bull bar as standard. Very noticeable in this design was that the ambulance body had a roof rack.

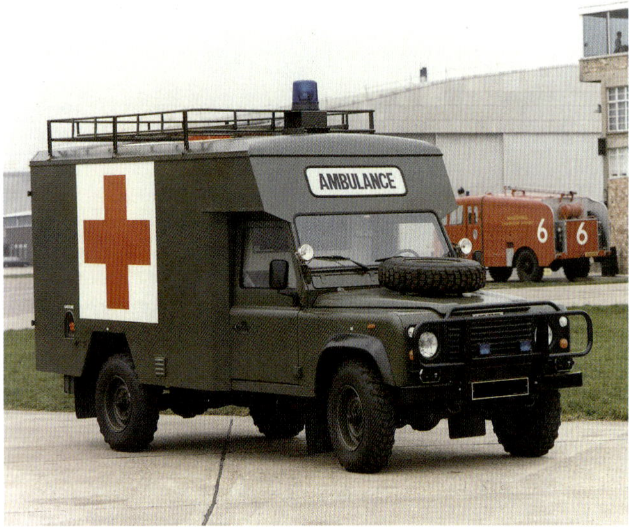

The UK MoD took a quantity of V8-engined 127 ambulances with Marshall bodies in the later 1980s.

A second batch of One Two Seven ambulances delivered to the British armed forces in the 1980s was not built by Marshalls, but its rear body design was essentially the same as the Marshalls type with accommodation for two or three stretchers. This second batch, again on V8-engined chassis and delivered in 1989 in the KJ registration series, was built by MMB (Macclesfield Motor Bodies) to a design by Locomotors of Andover. These later One Two Sevens were easily recognizable by their more angular Luton heads with a smaller 'Ambulance' sign. Some, and possibly all, also had a civilian-style smoked-glass window on the right-hand side, although there was no corresponding window on the left. Like the earlier Marshalls types, they all had a bull bar and a roof rack.

The UK MoD also took a large quantity of special One Two Seven models with V8 engines as tractors for the Rapier surface-to-air missile system that was typically used to defend airfields. The requirement came from the RAF Regiment, who specifically requested One Two Sevens to replace its 101 Forward Control Rapier tractors. The news reached Land Rover in March 1987 and Solihull prepared and delivered a 127 HCPU for MoD trials against a Ford-Iveco 4×4 truck.

The trials ended in mid-1989 and that July, Land Rover was awarded a contract to supply 214 vehicles for delivery over the next eight months or so. The first two were built in late October 1989 and were delivered to the Rapier Training Flight at RAF West Raynham in Norfolk for trials; the remaining vehicles were built from late November 1989 and the final examples were completed in February 1990. All had been delivered by the end of March 1990. Some were FFR models with a 12/24V electrical system and some were GS types with a standard 12V system. The two types were divided equally between LHD and RHD examples.

Although the 127 was a fully-developed vehicle, it did need to be adapted for its new role. Fitting a Warn 8274 winch at the front to assist in manoeuvring the Rapier trailers into place was easy enough, but design of the special back body to take the missile racks was subcontracted to Marshall Specialist Vehicles. The resulting pick-up type body

The 127 Rapier tractor, pictured on the left towing the missile launcher, was a special model developed for the RAF Regiment. The 127 Crew Cab behind, towing the Rapier optical tracking system, is a Land Rover military demonstrator and was not ordered by the UK armed forces.

was both longer and wider than anything else available on a Land Rover at the time. It had hatches in the sides ahead of the wheels to give access to equipment carried underneath the missile racks, together with a full-width tailgate. It also had a unique tall canvas tilt – drawn up by Graham Archer at Land Rover – with double supports at the front for additional rigidity. The cab roof also had a walkway so that Rapier crews could safely work on it when draping the vehicle with camouflage nets during an operational deployment.

One result of the extra-long rear overhang was that the RAF Regiment had to reconsider its arrangements for carrying a Rapier tractor (usually complete with trailer) underslung from a heavy-lift helicopter. The standard JATE attachment points at the rear of the chassis were now too far inboard and the lifting cable would have had to wrap around the rear of the body, so tending to pull the body from the chassis as the helicopter took the load. So a new system was devised, involving sling attachments bolted to the rear wheel hubs. The standard front attachments were unaffected and Rapier 127s would often be seen with the wooden beam across the top of the grille panel that protected the bonnet's front edge from damage during heliporting operations.

The major drawback was that Land Rover did not have a diesel engine powerful enough to meet the demands placed on a Rapier tractor and although the 200Tdi was just about to come on-stream, it was still an untried engine and the MoD did not want to take the risk. So to provide the power that the Rapier 127s needed, they were all fitted with the carburettor 3.5-litre V8.

In due course, this engine was to play a big part in the 127s' early withdrawal from service. As a petrol engine it presented logistical problems to a military force that was now committed to a single-fuel (diesel) policy. A secondary problem was that the new versions of the Rapier missile system that entered service in the early 1990s were bigger and heavier than the originals; the 127s needed yet more power. Then there was also the point that the Rapier detachments allocated to USAF bases in Britain were no longer needed after these closed down in the early 1990s. So by the middle of the decade, the 127 Rapier tractors had been relegated to support duties within the RAF Regiment; before long, they began to appear on the civilian market as surplus to requirements.

Ninety Variety

Despite the global trend towards larger vehicles, Land Rover was keen to secure military orders for the Ninety and by the mid-1990s had developed a 'gunship' variant in conjunction with Marshall Specialist Vehicles. These were equipped with a 106mm recoilless rifle in the load bed, with a split windscreen so that the long barrel of the gun could be rested on a front mount when the vehicle was in travelling mode. They were intended mainly as anti-tank weapons and the hope was that this derivative would appeal to those military forces who had taken the earlier 106mm 'gunship' version of the 88in Lightweight.

In search of new business, Land Rover did manage to persuade both the Singaporean and Malaysian armed forces to take trial examples of the Ninety Gunship with a view to replacing their ageing US-built M151 'Mutt' gunships. However, no orders followed. Some RHD examples were supplied to Cyprus, but the Ninety Gunship was withdrawn when it became obvious that the market was shrinking as the 106mm anti-tank rifle was gradually replaced by missile-based weapons.

The Ninety nevertheless did appear to have some potential as a missile carrier and Land Rover's Government and

ABOVE: **Exactly how the model helped to sell military Land Rovers is unexplained, but she did attract the photographer to this soft-top Ninety FFR demonstrator in the mid-1980s at a military exhibition.**

ABOVE: **Land Rover and Marshall's developed a Ninety Gunship in the later 1980s, but relatively few were sold.**

BELOW: **Land Rover promoted the Ninety as a compact weapons platform with demonstrators like this one in the late 1980s.**

LEFT: **During the Gulf War, a number of diesel Nineties were converted for use as convoy escort vehicles for the SAS. Few were ever photographed, but 62 KF 44 put in an appearance at a Dunsfold Collection Open Day.**

BELOW: **Otokar in Turkey began assembly of its own armoured patrol car on the One Ten chassis in the late 1980s. This was the production line in Istanbul.**

Military Operations division showed demonstrators modified to carry MILAN launchers at several military procurement events in the later 1980s. There was also some work to promote the Ninety as a desert patrol vehicle, but buyers preferred One Ten-based vehicles.

There was, however, an interesting exception. During the 1991 Gulf War, SAS units were deployed to act as convoy escorts in the desert, but there were not enough of their special One Ten V8 LRPVs to go round. So a number of standard-issue Nineties were commandeered and converted for the job. They were nicknamed 'Dinkies' – the D coming from their diesel engines and the rest of the name from the 'Pinky' nickname that was still being used for the One Tens.

Turkey (Otokar)

When the Turkish armed forces needed a fleet of light 4×4 vehicles, the deal that emerged involved a CKD assembly operation hosted by the established Otokar engineering company in Istanbul. Production began in 1987 and was initially of military vehicles only, but Otokar also assembled Land Rovers for local civilian use. The company also secured agreement to develop its own special military derivatives. All models were built on the One Ten and One Two Seven chassis, then subsequently on Defender 110 and Defender 130 chassis.

There was an armoured patrol vehicle on the 110 chassis from the late 1980s, similar in overall concept to the Shorland S55 of the time. However, in the early 1990s Otokar also developed the Akrep, a more specialized APC that barely resembled a Land Rover. This was offered with both V8 petrol and Tdi diesel engines. The Turkish armed forces took mainly 110s in a variety of guises from standard-looking Station Wagons to locally designed troop carriers and ambulances, but also bought ambulances on the One Two Seven and Defender 130 chassis.

In 1997, Otokar moved to new premises in Sakarya, Arifiye. From 2005, the company picked up the order to supply Defenders to the newly formed Iraqi armed forces, delivering both 110s and APCs.

ABOVE: **This RHD demonstrator was typical of the late military Ninety. It has a winch and special protective bumper at the front, plus rear bumperettes to avoid damage when push-starting with another vehicle.**

RIGHT: **The Dutch military trials at the end of the 1980s included an ambulance on the 127 chassis, powered by the Diesel Turbo engine. The ambulance body was by Marshall's of Cambridge.**

This was the core military One Ten by the end of the 1980s, in this case finished in a colour suggestive of desert use.

The Netherlands

The Royal Netherlands Air Force took a quantity of One Tens in 1988 as weapons platforms for air-defence duties, fitting them with weapons mounts in the rear to carry a .50 calibre machine gun. Some of these were petrol-powered, others had diesel engines and some of the petrol models were even equipped to run on Liquefied Petroleum Gas (LPG) – within NATO, something that was unique to the Dutch forces. The Dutch Marine Commandos also took a batch of 100 One Tens, equipping them with engine pre-heaters to ensure first-time starting in sub-zero temperatures.

The Royal Netherlands Army nevertheless lagged behind in ordering coil-sprung Land Rovers, mainly because of a series of problems that had occurred with the Series III models that had been bought in 1975. These problems had largely been caused by modifications the Dutch themselves had made and although Ninetys, One Tens and One Two Sevens with the Diesel Turbo engine were trialled in 1989–1990, the contract was awarded to the Mercedes-Benz G-Wagen, a vehicle that was increasingly encroaching on Land Rover's military sales in the 1980s and 1990s.

MRCV and SOV

By June 1992, Land Rover was able to claim that One Tens and their Defender 110 successors were in service with more than 150 armies around the world. The Ninety, however, had still not enjoyed great success as a military vehicle.

Yet in searching for a clearly defined role to suit the short-wheelbase model, Land Rover had already embarked on a development programme that would eventually lead to another successful derivative of the long-wheelbase type. For the Defender 90, Solihull's Government and Military Operations division had worked with Ricardo Specialist Vehicles to develop the 90 as a rapid-intervention vehicle with a ring-mount for weapons. It was announced as the MRCV (Multi-Role Combat Vehicle).

The MRCV was designed to be air-portable and had a modular design that enabled it to be reconfigured rapidly to suit different roles. The basic vehicle was a flatbed with integral roll cage and to this could be added various body configurations. In particular, the roll cage incorporated a ring-mount which enabled the vehicle to be used as a weapons platform for either a machine gun or MILAN missiles. There were large side and rear panniers for ammunition stowage or jerrycans.

The MRCV was exhibited widely but met with a fairly lukewarm reception, although in the later 1990s a further development of the design went to the Italian Special Forces as a Rapid Deployment Vehicle. Meanwhile, the US Rangers special forces unit had been impressed with the performance of the British Army's Land Rovers during the 1991 Gulf War. Like the British SAS, the Rangers were able to circumvent the normal military procurement process, so they approached Land Rover for a new vehicle that they needed quickly.

Land Rover responded in double-quick time with a Defender 110-based vehicle that shared many characteristics with the MRCV. It was again designed as a rapid-intervention vehicle; to suit both that role and US preferences, it was fitted with the 3.5-litre V8 petrol engine. The new vehicles for the Rangers were known as SOV (Special Operations Vehicle) types and Land Rover publicized their availability in June 1992 at the Eurosatory military exhibition in Paris.

Not surprisingly, the connection with the Rangers was not mentioned at the time. The show vehicle, which was probably the same one pictured in press handouts of the time, was finished to look like a rapid-attack vehicle for use in the desert. It was said to be available with either a 3.5-litre V8 petrol or a 200Tdi diesel engine. In practice, the US Rangers may have been the only customer. There were reports during 1992 that the first operational deployment of these vehicles was at the Olympic Games in Barcelona that summer, when there were concerns that US athletes might be the subject of a terrorist attack. The SOVs were reported as being painted matt black and concealed at various locations within Barcelona, ready to respond to any incident that might arise. Fortunately, none did.

CAV100: The 'Snatch' Vehicles

During the late 1980s, Courtaulds Aerospace developed a new lightweight armoured body for the long-wheelbase Land Rover and introduced it as the CAV100. The rear body 'tub' was made from a pressure-moulded laminate of glass-fibre and phenolic resin, a plastic-like substance that reduced the risk of spall inside the vehicle to almost nil. Though intended only to resist shrapnel and light arms fire, the CAV100's light weight made it far more manoeuvrable than earlier armoured 110s.

Among the customers for the CAV100 were the UK MoD, which ordered 1,000 examples with the 3.5-litre V8 engine in 1992. These were intended for use on peace-keeping duties in Northern Ireland, where their armour was adequate to protect against everyday threats. The vehicles gained the nickname of 'Snatch', because their speed and manoeuvrability allowed them to be used for operations where rioters could literally be 'snatched' off the streets of Belfast. Several examples later served with the UN peacekeeping force in the Balkans.

However, when British forces went into Afghanistan in 2001 and then Iraq in 2003, they took Snatch vehicles with them because they were the only armoured patrol vehicles available. Problems soon arose. One was heat inside the vehicles and a claimed 278 vehicles were returned to the UK in 2004–5 to have air-conditioning added. These were subsequently known as Snatch 1.5 types. Meanwhile, roadside bombs with far greater force than those seen in Northern Ireland were making the weaknesses of the armour glaringly apparent.

A further 300 vehicles were returned to the UK in two stages to have their bodies removed and put onto new XD-specification chassis with Tdi diesel engines by Hobson Industries Ltd; these were subsequently known as Snatch 2A and Snatch 2B types. At the end of 2010, as the first more heavily armoured patrol vehicles began to arrive, the early Snatch models were withdrawn, but the Snatch 2 types remained in service, relegated to duties for which their light armour was more appropriate.

Defender XD: The 'Wolf' military models

Although the Td5 engine became the core powerplant for Defenders after 1998, it was never militarized. For military customers, Land Rover kept the 300Tdi engine available, drawing on a large stockpile of engines it had created before 1998. The 300Tdi production lines were sold during 1998 to International Motors Ltd in Brazil and this company continued to supply Land Rover in the UK and elsewhere with 300Tdi engines as required. It also developed the engine further with a 2.8-litre swept volume and variable-geometry turbocharger, marketing this engine as the Power Stroke 2.8HS.

ABOVE: **The CAV100 was a lightly armoured vehicle that was ideal in the right circumstances.**

RIGHT: **The 'Snatch' – seen here as a revitalized Snatch 2 variant – earned itself an unfortunate reputation when it was used for tasks beyond its design capability.**

A British military winterized and waterproofed XD90 drives onto a landing craft during exercises.

This XD 110 belonging to the UK Military Police is wearing the special demountable hardtop designed for the Wolf vehicles.

That large stockpile of 300Tdi engines was needed for a big contract that Land Rover had been awarded in early 1996 to supply the UK MoD with a special set of Defender derivatives. This contract was for over 6,000 vehicles, which were actually built between 1998 and 2000. The original plan had envisaged a longer build period, but for various reasons Land Rover chose to complete production and deliveries early.

The story behind these very special Defenders had begun in November 1991, when the MoD had invited tenders for a large fleet of vehicles to replace its existing Land Rovers. Of the nineteen companies invited to submit proposals, sixteen declined. Only Land Rover, Iveco Ford and Steyr-Daimler-Puch agreed to provide test vehicles – and at this stage, only Land Rover was awarded a contract actually to do so.

The requirement was for vehicles in three categories and Land Rover submitted proposals for all three. In the TUL (Truck, Utility, Light) category, it proposed the Defender 90. For the TUM (Truck, Utility, Medium) contract, it proposed the Defender 110, and for the TUM(HD) contract that demanded a Heavy-Duty gun tractor it planned to develop an uprated Defender 110. Later, there would be a requirement for vehicles in a fourth category, too, when the MoD called for a fleet of new Battlefield Ambulances to replace its existing Land Rover-based models.

Land Rover spent around eighteen months developing Defenders to meet the MoD requirement. The project was code-named Wolf and the development programme was led by Paul Markwick. The upgraded Defenders were powered by militarized 200Tdi engines that breathed through dust filters on the front wings behind the wheel arches. The gearboxes were the still-secret R380 types that would enter production in early 1994, with a heavy-duty clutch and their own oil cooler. The Wolf prototypes also had the all-round disc brakes that would reach production models in 1993, plus power-assisted steering. They probably also had the new heavy-duty 165mm-wide wheel rims that GKN had designed specially for the purpose.

Some of the prototypes had 24V electrical systems and were configured as FFR types. No fully satisfactory militarized 200Tdi engine had yet been developed, so the large military generator had been bolted on where it would fit – which was too high to fit under the standard bonnet. So the 24V models had an offset bulge in their bonnets while Solihull's engineers developed a proper militarized version of the Gemini engine. This bulge not only looked odd; it also prevented the spare wheel from being mounted in its traditional military place on the bonnet. So it was mounted on a side-hinged tailgate instead. The TUM(HD) contenders also had a special wide back body designed to accommodate a standard NATO pallet, plus a taller rear roofline.

The vehicles delivered to the Defence Evaluation and Research Agency (DERA) for the military for trials in May 1993 had registrations in the KL series. Land Rover retained some for its own development work, giving them civilian registrations. However, the MoD's latest battlefield trials were very much tougher than Land Rover had anticipated. In September 1993, less than five months after the trials had begun, DERA called a halt to the trials because the Wolf prototypes were displaying major structural problems, together with axle, spring, damper and steering failures. The gearboxes were also overheating in extreme conditions. The MoD invited Solihull to put things right, giving its engineers twelve months in which to do the job.

The failure of that first batch of Wolf prototypes was so major that the only solution was a fundamental redesign of the vehicle. So the forty-five-strong project team set about redesigning their vehicles, referring to the second-phase prototypes as Wolf 2 types. Essentially, they embarked on

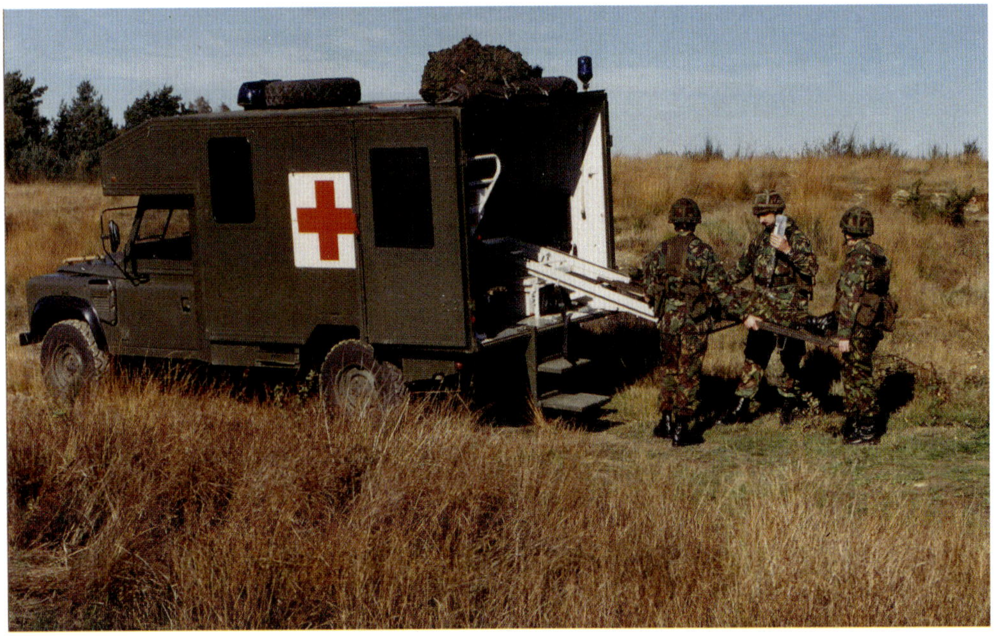

The XD 130 or Pulse ambulance was designed for the UK MoD.

a major programme of reinforcement, while further work on the R380 gearbox dealt with the overheating problem. When the Wolf 2 trials began in September 1994, it soon became apparent that Land Rover's work had been along the right lines. The new Goodyear G90 tyres fitted to all the trials vehicles had to be redesigned after the first ones proved prone to punctures, but otherwise there were only minor adjustments to be made. By May 1995, some nine months into the trials, the MoD pronounced itself to be satisfied with the TUL and TUM vehicles.

As for the TUM(HD) contract, however, Land Rover was unsuccessful. After the failure of the Wolf 1 vehicles, Steyr-Daimler-Puch had re-entered the trials and the TUM(HD) contract was eventually awarded to its Pinzgauer model. There was to be compensation elsewhere, though. Land Rover had worked with specialists Marshall's of Cambridge to develop a Battlefield Ambulance on a Defender 130 chassis that incorporated most of the Wolf 2 improvements, giving this project the code name of Pulse. In due course, the Pulse ambulance also satisfied the MoD trial requirements.

In September 1995, Land Rover displayed Wolf 2 trials vehicles at the British Army Equipment Exhibition and at the same time announced that the new model would be called the Defender XD when it entered production. That name stood for eXtra Duty. The official announcement of a production contract for £200 million-worth of Defender XDs followed in January 1996. The announcement of Land Rover's success with the Battlefield Ambulance contract was made some time later.

The differences between a Defender XD and its standard civilian equivalent were not immediately obvious, but the two vehicles were very different under the skin. The XD chassis had stronger cross-members and massive longitudinal reinforcement under the load bed. The axles had stronger half-shafts, four-pinion differentials and thicker casings. Both springs and dampers were upgraded and a guard plate protected the steering components.

The engine, which had also been fitted to the Wolf 2 trials vehicles, was now the 300Tdi diesel. As Chapter 7 explains, the front end of the Gemini engine had been redesigned to allow a 24V military generator to sit lower down and fit under a standard bonnet and the results had been fed into the refinement programme for the 200Tdi to produce an engine that suited all requirements. This was the Gemini 3 or 300Tdi.

The XD bodywork had also been heavily reinforced. There was a reinforced bulkhead, stronger sills and stronger top rails for the cab area, plus a strengthened radiator surround frame. Under the load floor, lateral reinforcement allowed a payload up to 20 per cent greater than that of the standard Defender. Fore and aft of the wheel arches were more reinforcing members in the sides of the load bed. The side-opening tailgate of the Wolf 1 prototypes had been abandoned in favour of a traditional drop-down type and there was a new spare-wheel mounting arrangement.

With the standard bonnet now a viable proposition, it was possible to carry the spare wheel on it and the Wolf 2 prototypes had the appropriate mounting. However, these Defender XDs were capable of higher speeds than their military predecessors over rough terrain and tests had shown that the weight of a spare wheel could damage the bonnet at such speeds. Reinforcing the bonnet made it too heavy for practical use, so Land Rover had developed a special spare-wheel mounting behind the door window on the passenger's side. The door mirror had an extended arm so that the driver could see past this obstruction.

Safety had also been a consideration in the design of these vehicles. Troops sitting in the back were provided with seat-belt anchorage points and over their heads was a demountable rollover cage that also gave extra strength to the body and acted as a support for the tilt cover. This cage made Wolf models taller than earlier military Land Rovers, although the extra height was not immediately obvious when the soft-top was fitted. It was clearly visible when the demountable hardtop was in place, though; the custom-made production type was made of GRP and incorporated a vision hatch in the cab roof, although trials vehicles had used a special metal hardtop.

The initial contracts called for 1,411 Defender 90s in the TUL class and 6,514 Defender 110s in the TUM class, plus around 800 Marshall-bodied Defender XD 130s. The first XD Defenders entered British military service on 1 April 1997 and by October 1998 the whole batch of nearly 8,000 vehicles had been constructed.

The GS, FFR and BFA (Battlefield Ambulance) variants were built first, with the last variant that Land Rover supplied to the MoD being a special derivative of the XD110 known as the WMIK (pronounced 'wimmick'). The letters stood for Weapons Mount Installation Kit and the demountable kit had been developed earlier in the second half of the 1990s by Land Rover in conjunction with Ricardo Special Vehicles. Earlier versions had been seen on the MRCV and SOV models. Its purpose was to give Land Rover a vehicle to suit the needs of rapid-reaction forces all over the world, while making that vehicle cost-effective. That was why the kit was demountable – when it was not needed, it could be removed and the host vehicle could be returned to its standard soft-top or hardtop configuration.

The core of the WMIK was a low-profile cage fitted to the rear of the vehicle, which both provided rollover protection for the crew and a stable ring-mount for a weapon. A hydraulically elevated observer's seat enabled that crew member to operate a machine gun mounted on the front bulkhead and the doors could be removed to speed access and exit for the crew. The WMIK kit could be fitted to a standard XD 110 in four hours and demounted in a similar time when no longer required. Its availability was announced in 1999, with the UK MoD initially taking a batch of 200 kits, which were shared between units of the Army and the Royal Marines. Since then, the basic concept of the WMIK has been further developed.

Once XD production for the UK MoD had been completed, Land Rover offered the Wolf variants to other military customers. Interest came from the Canadians, who examined a specially built Defender XD 110 Station Wagon with automatic transmission and investigated ways of armouring it. However, no order followed. During 2000, the Royal Netherlands Marine Corps invited Land Rover to tender for a contract to replace the One Tens they had been using since 1986. Although the current Dutch standard military 4×4 was the Mercedes-Benz G-Wagen, the Dutch were joint partners with the UK in the Amphibious Landing Force and it made sense to standardize vehicles and equipment as much as possible.

The Dutch waterproofing requirements were especially strict, and in developing vehicles to meet the Marines' requirements, Land Rover worked closely with Ricardo Special Vehicles of Shoreham. As the only European manufacturer to meet the criteria, it won the contract. The Dutch decided not to run extended trials, relying instead on data from the DERA trials of the Wolf prototypes, and deliveries of the Dutch XD 110s began in November 2001.

There were seventy-one of them, forty hardtops and thirty-one soft-tops, and all had the 24V FFR electrical specification. They were known as XD 110 HS/WW types, the additional letters standing for High-Specification, Waterproofed and Winterized. Their rollover cages incorporated a ring-mount for a machine-gun and the tilt covers of the soft-tops were specially made by KMA in Holland with a removable Kevlar lining that gave some protection against small-arms fire. During 2002, the Dutch took delivery of a second batch of vehicles, this time Marshall-bodied XD 130 ambulances. These were almost certainly the last Defender XDs to be built.

New Customers

With the end of the Cold War in 1990 and the gradual dismantling of barriers between the former western and eastern sectors of Europe, Land Rover was able to seize new opportunities to provide vehicles for the military forces of several former Soviet Bloc countries. The United Nations forces operating in Bosnia and other states of the former Yugoslavia also overwhelmingly used Land Rovers, a powerful demonstration of their value to military forces in the region.

Among those countries that became new customers for military Land Rovers were the Czech Republic and Croatia. The Czech tender went out in late 1994 and was awarded to Land Rover in June 1995; the first deliveries followed at the end of that year. They included 110 soft-tops and 110 Station Wagons, the latter particularly used by their Military Police. There were 90 Station Wagons and a SOV-like vehicle that was also mine-protected. Types on the 130 chassis included ambulances and van-bodied vehicles used among other things as mobile Nuclear, Biological, or Chemical (NBC) detection laboratories. By 2006, the Czechs also had some armoured 110s. The Croatian deliveries started rather later and included 110 Station Wagons for their Military Police.

In 2000, the Australian Defence Force took a quantity of Td5-powered window hardtops as an interim buy. This publicity picture shows one with all the Australian campaign medals that Land Rovers would have been awarded if they had been given to vehicles.

Land Rover also delivered military vehicles to a number of the western European countries during the 1990s. Portugal, which was a long-standing customer, took batches of Defender 90 and 110 Station Wagons. Belgium, another long-standing customer, took Defender 110 Station Wagons for a variety of duties that included Explosive Ordnance Disposal (EOD) work. Finland took a number of armoured CAV100s and a particularly important conquest sale was made to the Italian armed forces, following on from the sale of special Defender 90s to the Carabinieri in 1995.

Running orders continued to account for many military Land Rover sales. The Malaysian Army had been a regular customer since the early 1980s, with orders for more than 4,000 vehicles. Most of its orders were for 90 soft-tops, but there were also some 110 Station Wagons. Many of these vehicles were assembled in-territory from CKD kits. Brunei took regular deliveries during the first decade of the twenty-first century, mainly of 110 soft-tops, and the Indonesians took batches of 90s and 110s during the 1980s and 1990s, plus some XD 90 soft-tops for their special forces early in the new century. The Finnish armed forces also continued to take new Defenders.

Some of the developments from this period have still not been made public, but of interest was a pair of prototypes made in June 1994 in pursuit of a Singapore Army contract to replace their existing fleet of One Tens. They combined 300Tdi diesel engines with ZF 4-speed automatic gearboxes and had a soft-top configuration. One was shipped to Singapore for trials, while the other remained with the Military Engineering Division at Solihull. No order followed, but the episode does illustrate Land Rover's willingness to build vehicles more or less to order; the company had never built a Defender 300Tdi with automatic transmission before tendering for this contract.

One particularly interesting military order was placed by the Brazilian armed forces in 2003. It called for a fleet of Defender 90, 110 and 130 models in various configurations for the Army, Navy and Air Force to replace existing Toyota Bandeirante models. The vehicles were assembled in Brazil, where there was already a Defender CKD assembly operation (*see* Chapter 1).

It is not clear how large the order was, but by early 2004 the Army had received 870 vehicles, the Navy forty-five and the Air Force twenty. All were soft-top derivatives with rollover cages braced to the windscreen header rail. This configuration produced some unique-looking vehicles, such as four-door 130s with a canvas cab and separate soft-top for the load bed, 90s with tall window soft-tops and 110 ambulances with four canvas-panelled doors and a Station Wagon-like profile.

The Brazilian-built ambulances for that country's own Marines on Defender 110 chassis had canvas doors, as well as a canvas roof.

LEFT: **Also for the Brazilian Marines was this Defender 90, with heavy-duty wheel rims, canvas doors and canvas roof.**

ABOVE: **The Brazilian military adaptation of the new switch panel on 2002 and later models was interesting.**

Seen here during a handover ceremony to the Brazilian Army at Land Rio, the dealership in Rio de Janeiro, this is a Defender 130. Note again the canvas doors and roof; the windscreen could fold flat onto the bonnet – a feature not available on standard Land Rovers for many years.

ARMOURED MILITARY LAND ROVERS

Several varieties of armoured coil-sprung Land Rover have been manufactured by specialists for military use. The best-known is the Shorland armoured patrol car, manufactured by Short Brothers in Belfast and originally introduced as a conversion of the Series IIA model in 1965. The basic design was modified to suit the coil-sprung One Ten chassis and became available as a Shorland Series 5 from about 1983.

From 1985, Glover Webb gained Land Rover Approval for its rival product. This was called the Hornet and one of its key features was a low-profile turret that supposedly made the vehicle less conspicuous. Like the contemporary Shorland, the Hornet was based on a V8-engined One Ten chassis with the latest heavy-duty specification. Glover Webb also marketed the Hornet as an anti-terrorist vehicle, perhaps mindful of the fact that Schipol Airport in Holland had taken on a number of Shorlands for precisely that role.

RIGHT: **Short Bros modified its design with a sloping back panel because the original 'boot' had provided a place for rioters to lodge explosive devices. This example also has the special heavy-duty wheels fitted to later examples.**

BELOW: **Glover Webb offered its Hornet armoured vehicle for anti-terrorist duties and this demonstrator was configured for that role.**

ARMOURED MILITARY LAND ROVERS

The Shorland SB501 was an armoured troop transport. Note the heavy-duty wheels.

There were also Armoured Personnel Carriers from Shorts (the SB501) and from Glover Webb, the latter marketed as a riot-control vehicle. SMC Engineering of Bristol already had a three-axle chassis and on this it developed a formidable-looking armoured patrol car called the Hussar. This was supposedly bought by the armed forces of Sri Lanka. A two-axle version, based on a heavy-duty One Ten chassis, was called the Skirmisher and among its customers were the Pakistan border police in Peshawar. Other types were built by Otokar in Turkey (see Chapter 1).

ABOVE: **The Skirmisher was the two-axle equivalent of the Hotspur Hussar. This example was for the border police in north-west Pakistan.**

BELOW: **The Hotspur Hussar had only the barest of visual resemblances to the Land Rover at its core.**

Refurbishment Programmes

A marked trend during the first decade of the twentieth century was for military forces to refurbish and upgrade their existing fleets of coil-sprung Land Rovers rather than to buy new. One reason for this was undoubtedly the question mark hanging over the long-term future of the Defender, but another was the increasing need to save money. The Australian Defence Force was one of the earliest to embark on a refurbishment programme as it tackled its Perentie 6×6 fleet in order to extend its service life.

Even Land Rover jumped onto the bandwagon in 2007, developing a 6×6 prototype in anticipation of a UK MoD requirement for a 3,000–4,000kg (6,615–8,820lb) vehicle. Longer-term, Solihull expected the MoD to replace its older One Tens and wanted to be ready with ideas for when it did. Working with Ricardo Special Vehicles, the company bought four Perentie 6×6 logistics vehicles that had formerly been in service in Oman and built two of them up to meet the likely requirement. The other two were retained for spares, as the Perentie 6×6 was no longer in production. Both were extensively refurbished and upgraded with the latest 2.4-litre Puma diesel engine and 6-speed gearbox.

However, that project came to nothing. The UK MoD decided that it would be more cost-effective to refurbish its existing Land Rovers, so at the end of the decade embarked on two separate programmes. Project Tithonus was described as a mid-life improvement programme and dealt with 1,284 One Tens from the second half of the 1980s, ending in mid-2010. The refurbishment was carried out at Catterick, Donnington and Stirling and was expected to extend the vehicles' service life by ten years. The main improvements were to safety (as newly required by the Health and Safety Executive), plus the vehicles were given rollover protection, better seating, rear seat belts and a noise and vibration reduction package.

The second programme followed the first in 2010, and was called Project Wolf Remus. Its purpose was to extend the service life of around 6250 XD 110 vehicles from the originally projected date of 2017 to 2030. The measures taken included fitting a new noise and vibration-reduction package. From 2009, UK MoD also looked at converting some of its XD 130 Battlefield Ambulances into mortar and specialist team vehicles, using a double-cab configuration with a tall roof that covered an internal roll cage.

The tradition continues: this is a military version of the Puma-engined Defender with the Utility Station Wagon body, for an unidentified end-user.

APPENDIX

THE COIL-SPRUNG LAND ROVERS: A QUICK GUIDE

The thirty-year story of the coil-sprung Land Rovers is a complicated one, but this guide simplifies it by showing the major development milestones.

Model Year	Drivetrain, Structure and New Models	Exterior	Interior
1983	2.3P, 2.3D, 3.5V8	Limestone upper panels; Limestone side tapes for County; galvanized bumper	Black vinyl; County with brown cloth
1984	Ninety; 2.5 D		
1985	LT85 for V8s	Winding windows in two-piece doors; body-colour upper panels;	Ivory White for roofs; County tapes with trailing dashes; Matt Stone soft-tops
1986	2.5 P		
1987	Add Diesel Turbo; 124bhp V8 for Europe	County with body-colour eyebrows and hockey-stick tapes; rear vents	
1988		Black bumper; County with grille painted to match body and angled side tapes; sun-hatch option	Grey vinyl or (for County) grey cloth; black door trims
1989		Ribless roof; grey side tapes for standard models	
1990		90, 110 and 127 grille badges; rear vents deleted	
Defender			
1991	200Tdi plus LT77	Green and red Defender side tapes; 'strobe' County tapes	County to Moorland Grey cloth
1992	LT77S	Grey and yellow Defender side tapes	

Model Year	Drivetrain, Structure and New Models	Exterior	Interior
1993	NAS 110	Spare wheel cover disc; first metallic paints; non-County tapes deleted mid-year; windscreen hinges deleted	
1994	NAS 90; rear discs	First alloy wheels for 90	
1995	300Tdi plus R380		
1996		First metallic and micatallic paints for County	
1997	Dished alloys for 110		Twill PVC; Raleigh cloth for County; Dark Granite door trims
1998	(V8 50)		
1999	Td5 plus R380	Body-colour roof for County; no side tapes	Dark Granite Twill PVC; Dark Granite Techno cloth for County
2000	Double Cab (May)	Body-colour eyebrows and roofs on truck cabs	
2002		Steel doors; 90 bulkhead deleted SW windscreen option	Central switch panel; heated seats and on
2003	XS trim level introduced		
2007	Puma 2.4 engine plus MT82; Utility SW added	Bulge added to bonnet	New dashboard; forward-facing rearmost seats
2012	Puma 2.2 engine plus MT82		

INDEX